Forestry: A Sustainable Perspective

Forestry: A Sustainable Perspective

Edited by

Timothy Armstrong

Larsen & Keller
www.larsen-keller.com

Forestry: A Sustainable Perspective
Edited by Timothy Armstrong
ISBN: 978-1-63549-127-2 (Hardback)

Larsen & Keller
Published by Larsen and Keller Education,
5 Penn Plaza,
19th Floor,
New York, NY 10001, USA

Cataloging-in-Publication Data

Forestry : a sustainable perspective / edited by Timothy Armstrong.
p. cm.
Includes bibliographical references and index.
ISBN 978-1-63549-127-2
1. Forests and forestry. 2. Sustainable forestry.
3. Forest management. I. Armstrong, Timothy.
SD131 .F67 2017
634.9--dc23

The publisher's policy is to use permanent paper from mills that operate a sustainable forestry policy. Furthermore, the publisher ensures that the text paper and cover boards used have met acceptable environmental accreditation standards.

Printed and bound in China.

For more information regarding Larsen and Keller Education and its products, please visit the publisher's website www.larsen-keller.com

Table of Contents

Preface

The book aims to shed light on some of the unexplored aspects of forestry. It picks up individual branches and explains their need and the contribution they make in the context of the growth of this field. Forestry refers to the practice of maintaining, conserving, repairing, protecting and using the forest and its resources in such a way that it fulfills the current requirements and is conserved for the future generations. This book attempts to understand the multiple branches that fall under forestry management and how such concepts have practical applications. This text is a valuable compilation of topics, ranging from the basic to the most complex theories and principles in the field of forestry. Some of the diverse topics covered in it address the varied branches that fall under this category. The textbook will serve as a valuable source of reference to students and for those interested in this field.

A foreword of all Chapters of the book is provided below:

Chapter 1 - Forestry practices have been a part of human communities since the early past and they have continued to be so up to the present. Forestry is an important part of our culture and economy. The challenge of forestry is to be sustainable in the 21st century. This chapter on forestry offers an insightful focus, keeping in mind the complex subject matter; **Chapter 2** - Forestry cannot be defined as one single system as it requires planned tree-felling practices as well as correct cultivation and afforestation techniques. Forestry is dependent on many other practices that have evolved over the years. Forestry is an interdisciplinary subject. It spreads to other fields as well. This chapter will provide a glimpse of diverse aspects of forestry briefly; **Chapter 3** - One of the most important practices involved in forestry is logging and tree cutting for timber. Subjects involved in this chapter include various types of logging, rubber tapping, clear cutting and plant nurseries. These practices maintain the balance between timber being felled and new saplings that are being cultivated. The chapter strategically encompasses and incorporates the major components and key concepts of forestry, providing a complete understanding; **Chapter 4** - Forestry was born out of the conservation movement in the late 19th and early 20th century. It has seen various avatars in the hands of professional foresters and has been the source of wealth for many nations. Recent events of climate change, global warming and high concentrations of the urban environment have all led to concerted efforts of sustainable forestry practices. This chapter lists some of them such as landscape ecology and forest dynamics studies; **Chapter 5** - Forest produce comprise of an important part of local economies as they can provide livelihood and resource. Timber is also treated before it is processed as a commodity. Some of the forest resource processing techniques are woodworking, wood preservation and paperboard making; **Chapter 6** - Forestry requires systematic data collection on tree-felling, afforestation as well as the occurrence of natural hazards such as landslides and wildfires. Some methods of data collection include basal area, analog forestry and geographic information systems. The chapter serves as a source to understand the major categories related to forestry; **Chapter 7** - In order to ensure legal and correct methods of tree-felling and timber production, organizations have created measures of standardisation that reflect the quality of the product as well as safe environmental conditions. Some measures include certified wood, old-growth forests and environmental audits. This chapter is a compilation of the various branches of forestry that form an integral part of the broader subject matter.

I would like to thank the entire editorial team who made sincere efforts for this book and my family who supported me in my efforts of working on this book. I take this opportunity to thank all those who have been a guiding force throughout my life.

Editor

1

Introduction to Forestry

Forestry practices have been a part of human communities since the early past and they have continued to be so up to the present. Forestry is an important part of our culture and economy. The challenge of forestry is to be sustainable in the 21st century. This chapter on forestry offers an insightful focus, keeping in mind the complex subject matter.

Forestry is the science and craft of creating, managing, using, conserving, and repairing forests and associated resources to meet desired goals, needs, and values for human and environment benefits. Forestry is practiced in plantations and natural stands. The science of forestry has elements that belong to the biological, physical, social, political and managerial sciences.

Forestry work in Austria

Modern forestry generally embraces a broad range of concerns, in what is known as multiple-use management, including the provision of timber, fuel wood, wildlife habitat, natural water quality management, recreation, landscape and community protection, employment, aesthetically appealing landscapes, biodiversity management, watershed management, erosion control, and preserving forests as 'sinks' for atmospheric carbon dioxide. A practitioner of forestry is known as a forester. Other terms are used a verderer and a silviculturalist being common ones. Silviculture is narrower than forestry, being concerned only with forest plants, but is often used synonymously with forestry.

Forest ecosystems have come to be seen as the most important component of the biosphere, and forestry has emerged as a vital applied science, craft, and technology.

Forestry is an important economic segment in various industrial countries. For example, in Germany, forests cover nearly a third of the land area, wood is the most important renewable resource, and forestry supports more than a million jobs and about billion in yearly turnover.

A deciduous beech forest in Slovenia

History

Background

The preindustrial age has been dubbed by Werner Sombart and others as the 'wooden age', as timber and firewood were the basic resources for energy, construction and housing. The development of modern forestry is closely connected with the rise of capitalism, economy as a science and varying notions of land use and property.

Roman Latifundiae, large agricultural estates, were quite successful in maintaining the large supply of wood that was necessary for the Roman Empire. Large deforestations came with respectively after the decline of the Romans. However already in the 5th century, monks in the then Byzantine Romagna on the Adriatic coast, were able to establish stone pine plantations to provide fuelwood and food. This was the beginning of the massive forest mentioned by Dante Alighieri in his 1308 poem Divine Comedy.

Similar sustainable formal forestry practices were developed by the Visigoths in the 7th century when, faced with the ever increasing shortage of wood, they instituted a code concerned with the preservation of oak and pine forests. The use and management of many forest resources has a long history in China as well, dating back to the Han Dynasty and taking place under the landowning gentry. A similar approach was used in Japan. It was also later written about by the Ming Dynasty Chinese scholar Xu Guangqi (1562–1633).

In Europe, land usage rights in medieval and early modern times allowed different users to access forests and pastures. Plant litter and resin extraction were important, as pitch (resin) was essential for the caulking of ships, falking and hunting rights, firewood and building, timber gathering in wood pastures, and for grazing animals in forests. The notion of "commons" (German "Allmende") refers to the underlying traditional legal term of common land. The idea of enclosed private property came about during modern times. However, most hunting rights were retained by members of the nobility which preserved the right of the nobility to access and use common land for recreation, like fox hunting.

Early Modern Forestry Development

Timber harvesting, as here in Finland, is a common component of forestry

Systematic management of forests for a sustainable yield of timber is said to have begun in the German states in the 14th century, e.g. in Nuremberg, and in 16th-century Japan. Typically, a forest was divided into specific sections and mapped; the harvest of timber was planned with an eye to regeneration. As timber rafting allowed for connecting large continental forests, as in south western Germany, via Main, Neckar, Danube and Rhine with the coastal cities and states, early modern forestry and remote trading were closely connected. Large firs in the black forest were called „Holländer", as they were traded to the Dutch ship yards. Large timber rafts on the Rhine were 200 to 400m in length, 40m in width and consisted of several thousand logs. The crew consisted of 400 to 500 men, including shelter, bakeries, ovens and livestock stables. Timber rafting infrastructure allowed for large interconnected networks all over continental Europe and is still of importance in Finland.

Starting with the sixteenth century, enhanced world maritime trade, a boom in housing construction in Europe and the success and further Berggeschrey (rushes) of the mining industry increased timber consumption sharply. The notion of 'Nachhaltigkeit', sustainability in forestry, is closely connected to the work of Hans Carl von Carlowitz (1645–1714), a mining administrator in Saxony. His book *Sylvicultura oeconomica, oder haußwirthliche Nachricht und Naturmäßige Anweisung zur wilden Baum-Zucht* (1713) was the first comprehensive treatise about sustainable yield forestry. In the UK, and to an extend in continental Europe, the enclosure movement and the clearances favored strictly enclosed private property. The Agrarian reformers, early economic writers and scientists tried to get rid of the traditional commons. At the time, an alleged tragedy of the commons together with fears of a Holznot, an imminent wood shortage played a watershed role in the controversies about cooperative land use patterns.

Hans Carl von Carlowitz

The practice of establishing tree plantations in the British Isles was promoted by John Evelyn, though it had already acquired some popularity. Louis XIV's minister Jean-Baptiste Colbert's oak Forest of Tronçais, planted for the future use of the French Navy, matured as expected in the mid-19th century: "Colbert had thought of everything except the steamship," Fernand Braudel observed. In parallel, schools of forestry were established beginning in the late 18th century in Hesse, Russia, Austria-Hungary, Sweden, France and elsewhere in Europe.

Forest Conservation and Early Globalization

During the late 19th and early 20th centuries, forest preservation programs were established in British India, the United States, and Europe. Many foresters were either from continental Europe (like Sir Dietrich Brandis), or educated there (like Gifford Pinchot). Sir Dietrich Brandis is considered the father of tropical forestry, European concepts and practices had to be adapted in tropical and semi arid climate zones. The development of plantation forestry was one of the (controversial) answers to the specific challenges in the tropical colonies. The enactment and evolution of forest laws and binding regulations occurred in most Western nations in the 20th century in response to growing conservation concerns and the increasing technological capacity of logging companies. Tropical forestry is a separate branch of forestry which deals mainly with equatorial forests that yield woods such as teak and mahogany.

Mechanization

Forestry mechanization was always in close connection to metal working and the development of mechanical tools to cut and transport timber to its destination. Rafting belongs to the earliest means of transport. Steel saws came up in the 15th century. The 19th century widely increased the availability of steel for whipsaws and introduced Forest railways and railways in general for transport and as forestry customer. Further human induced changes, however, came since World War II, respectively in line with the '1950s-syndrome'. The first portable chainsaw was invented in 1918 in Canada, but large impact of mechanization in forestry started after World War II. Forestry Harvesters are among the most recent developments. Although drones, planes, laser scanning, satellites and robots also play a part in forestry.

Forestry Today

Today a strong body of research exists regarding the management of forest ecosystems and genetic improvement of tree species and varieties. Forestry also includes the development of better methods for the planting, protecting, thinning, controlled burning, felling, extracting, and processing of timber. One of the applications of modern forestry is reforestation, in which trees are planted and tended in a given area.

Trees provide numerous environmental, social and economic benefits for people. In many regions the forest industry is of major ecological, economic, and social importance. Third-party certification systems that provide independent verification of sound forest stewardship and sustainable forestry have become commonplace in many areas since the 1990s. These certification systems were developed as a response to criticism of some forestry practices, particularly deforestation in less developed regions along with concerns over resource management in the developed world. Some certification systems are criticized for primarily acting as marketing tools and lacking in their claimed independence.

In topographically severe forested terrain, proper forestry is important for the prevention or minimization of serious soil erosion or even landslides. In areas with a high potential for landslides, forests can stabilize soils and prevent property damage or loss, human injury, or loss of life.

A modern sawmill

Public perception of forest management has become controversial, with growing public concern over perceived mismanagement of the forest and increasing demands that forest land be managed for uses other than pure timber production, for example, indigenous rights, recreation, watershed management, and preservation of wilderness, waterways and wildlife habitat. Sharp disagreements over the role of forest fires, logging, motorized recreation and other issues drives debate while the public demand for wood products continues to increase.

Foresters

Foresters of UACh in the Valdivian forests of San Pablo de Tregua, Chile

Foresters work for the timber industry, government agencies, conservation groups, local authorities, urban parks boards, citizens' associations, and private landowners. The forestry profession includes a wide diversity of jobs, with educational requirements ranging from college bachelor's degrees to PhDs for highly specialized work. Industrial foresters plan forest regeneration starting with careful harvesting. Urban foresters manage trees in urban green spaces. Foresters work in tree nurseries growing seedlings for woodland creation or regeneration projects. Foresters improve tree genetics. Forest engineers develop new building systems. Professional foresters measure and model the growth of forests with tools like geographic information systems. Foresters may combat insect infestation, disease, forest and grassland wildfire, but increasingly allow these natural aspects of forest ecosystems to run their course when the likelihood of epidemics or risk of life or property are low. Increasingly, foresters participate in wildlife conservation planning and watershed protection. Foresters have been mainly concerned with timber management, especially reforestation, maintaining forests at prime conditions, and fire control.

Forestry Plans

Foresters develop and implement forest management plans relying on mapped resource inventories showing an area's topographical features as well as its distribution of trees (by species) and other plant cover. Plans also include landowner objectives, roads, culverts, proximity to human habitation, water features and hydrological conditions, and soils information. Forest management plans typically include recommended silvicultural treatments and a timetable for their implementation. Application of digital maps in Geographic Informations systems (GIS) that extracts and integrates different information about forest terrains, soil type and tree covers, etc. using, e.g. laser scanning, enhances forest management plans in modern systems.

Forest management plans include recommendations to achieve the landowner's objectives and desired future condition for the property subject to ecological, financial, logistical (e.g. access to resources), and other constraints. On some properties, plans focus on producing quality wood products for processing or sale. Hence, tree species, quantity, and form, all central to the value of harvested products quality and quantity, tend to be important components of silvicultural plans.

Good management plans include consideration of future conditions of the stand after any recommended harvests treatments, including future treatments (particularly in intermediate stand treatments), and plans for natural or artificial regeneration after final harvests.

The objectives of landowners and leaseholders influence plans for harvest and subsequent site treatment. In Britain, plans featuring "good forestry practice" must always consider the needs of other stakeholders such as nearby communities or rural residents living within or adjacent to woodland areas. Foresters consider tree felling and environmental legislation when developing plans. Plans instruct the sustainable harvesting and replacement of trees. They indicate whether road building or other forest engineering operations are required.

Agriculture and forest leaders are also trying to understand how the climate change legislation will affect what they do. The information gathered will provide the data that will determine the role of agriculture and forestry in a new climate change regulatory system.

Forestry as a Science

Over the past centuries, forestry was regarded as a separate science. With the rise of ecology and environmental science, there has been a reordering in the applied sciences. In line with this view, forestry is one of three primary land-use sciences. The other two are agriculture and agroforestry. Under these headings, the fundamentals behind the management of natural forests comes by way of natural ecology. Forests or tree plantations, those whose primary purpose is the extraction of forest products, are planned and managed utilizing a mix of ecological and agroecological principles.

Education

History of Forestry Education

The first dedicated forestry school was established by Georg Ludwig Hartig at Hungen in the Wetterau, Hesse, in 1787, though forestry had been taught earlier in central Europe, including at the University of Giessen, in Hesse-Darmstadt.

In Spain, the first forestry school was the Forest Engineering School of Madrid (Escuela Técnica Superior de Ingenieros de Montes), founded in 1844.

The first in North America, the Biltmore Forest School was established near Asheville, North Carolina, by Carl A. Schenck on September 1, 1898, on the grounds of George W. Vanderbilt's Biltmore Estate. Another early school was the New York State College of Forestry, established at Cornell University just a few weeks later, in September 1898. Early 19th century North American foresters went to Germany to study forestry. Some early German foresters also emigrated to North America.

In South America the first forestry school was established in Brazil, in Viçosa, Minas Gerais, in 1962, and moved the next year to become a faculty at the Federal University of Paraná, in Curitiba.

Forestry Education Today

Today, forestry education typically includes training in general biology, botany, genetics, soil science, climatology, hydrology, economics and forest management. Education in the basics of sociology and political science is often considered an advantage.

Prescribed burning is used by foresters to reduce fuel loads

In India, forestry education is imparted in the agricultural universities and in Forest Research Institutes (deemed universities). Four year degree programmes are conducted in these universities at the undergraduate level. Masters and Doctorate degrees are also available in these universities.

In the United States, postsecondary forestry education leading to a Bachelor's degree or Master's degree is accredited by the Society of American Foresters.

In Canada the Canadian Institute of Forestry awards silver rings to graduates from accredited university BSc programs, as well as college and technical programs.

In many European countries, training in forestry is made in accordance with requirements of the Bologna Process and the European Higher Education Area.

The International Union of Forest Research Organizations is the only international organization that coordinates forest science efforts world-wide.

References

- T. Mirov, Nicholas; Hasbrouck, Jean (1976). "6". The story of pines. Bloomington and London: Indiana University Press. p. 111. ISBN 0-253-35462-5.
- Braudel, Fernand (1979). The Wheels of Commerce: Civilization and Capitalism: 15th-18th Century (Volume II). University of California Press. p. 240. ISBN 978-0-520-08115-4.
- Petru-Ioan Becheru (Aug 2012). "Revista pădurilor online". Rev. pădur. (in Romanian). 127 (4): 46–53. ISSN 1583-7890. 16819. Retrieved 2012-10-21.
- Victor Giurgiu (Nov 2011). "Revista pădurilor (Journal of forests) 125 years of existence". Rev. pădur. 126 (6): 3–7. ISSN 1583-7890. Retrieved 2012-04-06.

2

Diverse Aspects of Forestry

Forestry cannot be defined as one single system as it requires planned tree-felling practices as well as correct cultivation and afforestation techniques. Forestry is dependent on many other practices that have evolved over the years. Forestry is an interdisciplinary subject. It spreads to other fields as well. This chapter will provide a glimpse of diverse aspects of forestry briefly.

Silviculture

Silviculture is the practice of controlling the establishment, growth, composition, health, and quality of forests to meet diverse needs and values.

The name comes from the Latin *silvi-* (forest) + culture (as in growing). The study of forests and woods is termed silvology. Silviculture also focuses on making sure that the treatment(s) of forest stands are used to preserve and to better their productivity.

Generally, silviculture is the science and art of growing and tending forest crops, based on a knowledge of silvics, i.e., the study of the life history and general characteristics of forest trees and stands, with particular reference to locality factors. More particularly, silviculture is the theory and practice of controlling the establishment, composition, constitution, and growth of forests. No matter how forestry as a science is constituted, the kernel of the business of forestry is silviculture, as it includes direct action in the forest, and in it all economic objectives and technical considerations ultimately converge. The kernel of silviculture is regeneration.

An employee of the White Mountain Apache Tribe forestry department uses a hoedad to dig a hole for a ponderosa pine seedling in the shade of a burned tree stump, which protects the seedling from windy conditions that dry the trees

Suggestions for how best to go about the job, presented by Jeglum et al. (2003), though aimed primarily at the boreal forest in Ontario, merit wider consideration. The 110-page publication describes Best Management Practices, first by general principles, then by sensitive sites. Illustrations are plentiful and are well chosen to complement this excellent text.

To some the distinction between forestry and silviculture is that silviculture is applied at the stand level and forestry is broader. For example, John D. Matthews says "complete regimes for regenerating, tending, and harvesting forests" are called "silvicultural systems".

So adaptive management is required for silviculture, whereas forestry can be natural, conserved land without a stand level treatment being applied. A common taxonomy divides silviculture into regenerating, tending and harvesting techniques.

Silvicultural Systems

The origin of forestry in German speaking Europe has defined silvicultural systems broadly as Hochwald - High forest, Mittlewald - Coppice with standards (and Compound coppice), Short rotation coppice and Niederwald - Coppice. There are other systems too. These varied silvicultural systems include several harvesting methods, which are often wrongly said to be a silvicultural systems, but may also be called as rejuvenating or regenerating method depending on the purpose.

Hochwald is further subdivided in German into: Altersklassenwald - Age class forest which includes:

- *even aged forestry*
    - - Kahlschlag - Clear cutting
    - - Schirmschlag - Shelterwood cutting
    - - Seed-tree method
- *un-even aged forestry*
    - - Femelschlag (Femelschlagverfahren) - The Femel selection cutting (Group selection cutting),
    - - Saumschlag (Saumfemelschlagbetrieb) - Strip selection cutting (strip-and-group felling system)
    - - Schirmkeilschlag - Shelterwood wedge cutting
    - - Mischformen - mixed form regeneration methods.

Dauerwald - Continuous cover forestry, in which are found:

- *un-even aged forestry*
    - - Plenterwald - Selection forest
    - - Zielstärkennutzung - Target diameter harvesting.

From these names the impression is given that these are neatly defined systems, but in practice there are a variety of variations within these harvesting methods according to local ecology and site conditions. While an individual location can be identified where it could be said to be the archetypal form of harvesting technique (they all originated somewhere with a particular forester, and have been described in the scientific literature), and broad generalizations can be made these are merely rules of thumb rather than strict blue prints on how techniques might be applied. This misunderstanding has meant that many older English textbooks did not capture the true complexity of silviculture as practiced where it originated in Mitteleuropa.

This silviculture was culturally predicated on wood production in temperate and boreal climates and did not deal with tropical forestry. The misapplication of this philosophy to those tropical forests has been problematic. There is also an alternative silvicultural tradition which developed in Japan and thus created a different biocultural landscape called satoyama.

After harvesting comes regeneration, which may be split into Natural and Artificial and tending which includes release treatments, pruning, thinning and intermediate treatments. As it can be imagined it maybe that any of these 3 phases (harvesting - regeneration - tending) may happen at the same time, within a stand, depending on the goal for that particular stand.

Regeneration

Regeneration is basic to the continuation of forested, as well as to the afforestation of treeless land. Regeneration can take place through self-sown seed ("natural regeneration"), by artificially sown seed, or by planted seedlings. In whichever case, the performance of regeneration depends on its growth potential and the degree to which its environment allows the potential to be expressed. Seed, of course, is needed for all regeneration modes, both for natural or artificial sowing and for raising planting stock in a nursery.

Natural regeneration is:

"Human-assisted natural regeneration" means establishment of a forest age class from natural seeding or sprouting in an area after harvesting in that area through selection cutting, shelter (or seed-tree) harvest, soil preparation, or restricting the size of a clear-cut stand to secure natural regeneration from the surrounding trees.

The process of natural regeneration involves the renewal of forests by means of self-sown seeds, root suckers, or coppicing. In natural forests, conifers rely almost entirely on regeneration through seed. Most of the broadleaves, however, are able to regenerate by the means of emergence of shoots from stumps (coppice) and broken stems.

Seedbed Requirements

Any seed, self-sown or artificially applied, requires a seedbed suitable for securing germination.

In order to germinate, a seed requires suitable conditions of temperature, moisture, and aeration. For seeds of many species, light is also necessary, and facilitates the germination of seeds in other species, but spruces are not exacting in their light requirements, and will germinate without light. White spruce seed germinated at 35 °F (1.7 °C) and 40 °F (4.4 °C) after continuous stratification for one year or longer and developed radicles < 6 cm long in the cold room. When exposed to light, those germinants developed chlorophyll and were normally phototropic with continued elongation.

For survival in the short and medium terms, a germinant needs: a continuing supply of moisture; freedom from lethal temperature; enough light to generate sufficient photosynthate to support respiration and growth, but not enough to generate lethal stress in the seedling; freedom from browsers, tramplers, and pathogens; and a stable root system. Shade is very important to the survival of young seedlings. In the longer term, there must be an adequate supply of essential nutrients and an absence of smothering.

In undisturbed forest, decayed windfallen stemwood provides the most favourable seedbed for germination and survival, moisture supply being dependable, and the elevation of seedlings somewhat above the general level of the forest floor reduces the danger of smothering by leaves and snow-pressed minor vegetation; nor is such a microsite likely to be subject to flooding. Advantages conferred by those microsites include: more light, higher temperatures in the rooting zone, and better mycorrhizal development. In a survey in the Porcupine Hills, Manitoba, 90% of all spruce seedlings were rooted in rotten wood.

Mineral soil seedbeds are more receptive than the undisturbed forest floor, and are generally moister and more readily rewetted than the organic forest floor. However, exposed mineral soil, much more so than organic-surfaced soil, is subject to frost heaving and shrinkage during drought. The forces generated in soil by frost or drought are quite enough to break roots.

The range of microsites occurring on the forest floor can be broadened, and their frequency and distribution influenced by site preparation. Each microsite has its own microclimate. Microclimates near the ground are better characterized by vapour pressure deficit and net incident radiation, rather than the standard measurements of air temperature, precipitation, and wind pattern.

Aspect is an important component of microclimate, especially in relation to temperature and moisture regimes. Germination and seedling establishment of Engelmann spruce were much better on north than on south aspect seedbeds in the Fraser Experimental Forest, Colorado; the ratios of seeds to 5-year-old seedlings were determined as 32:1, 76:1, and 72:1 on north aspect bladed-shaded, bladed-unshaded, and undisturbed-shaded seedbeds, respectively. Clearcut openings of 1.2 ha to 2.0 ha adjacent to an adequate seed source, and not more than 6 tree heights wide, could be expected to secure acceptable regeneration (4,900, 5-year-old trees/ha), whereas on undisturbed-unshaded north aspects, and on all seedbed treatments tested on south aspects, seed to seedling ratios were so high that the restocking of any clearcut opening would be questionable.

At least 7 variable factors may influence seed germination: (1) seed characteristics, (2) light, (3) oxygen, (4) soil reaction (ph), (5) temperature, (6) moisture, and (7) seed enemies. Moisture and temperature are the most influential, and both are affected by exposure. The difficulty of securing natural regeneration of Norway spruce and Scots pine in northern Europe led to the adoption of various forms of reproduction cuttings that provided partial shade or protection to seedlings from hot sun and wind. The main objective of echeloned strips or border-cuttings with northeast exposure was to protect regeneration from overheating, and was originated in Germany and deployed successfully by Alarik (1925) and others in Sweden. On south and west exposures, direct insolation and heat reflected from tree trunks often result in temperatures lethal to young seedlings, as well as desiccation of the surface soil, which inhibits germination. The sun is less injurious on eastern exposures because of the lower temperature in the early morning, related to higher humidity and presence of dew.

Baldwin (1933), after noting that summer temperatures in North America are often higher than those in places where border-cuttings have been found useful, reported the results of a survey of regeneration in a stand of red spruce plus scattered white spruce that had been isolated by clearcutting on all sides, so furnishing an opportunity for observing regeneration on different exposures in this old-field stand at Dummer, New Hampshire. The regeneration included a surprisingly large number of balsam fir seedlings from the 5% stand component of that species. The maximum density of spruce regeneration, determined 4 rods (20 m) inside from the edge of the stand on a North 20°E exposure, was 600,000/ha, with almost 100,000 balsam fir seedlings.

A prepared seedbed remains receptive for a relatively short period, seldom as long as 5 years, sometimes as short as 3 years. Seedbed receptivity on moist, fertile sites decreases with particular rapidity, and especially on such sites, seedbed preparation should be scheduled to take advantage of good seed years. In poor seed years, site preparation can be carried out on mesic and drier sites with more chance of success, because of the generally longer receptivity of seedbeds there than those on moister sites. Although an indifferent seed year can suffice if seed distribution is good and environmental conditions favourable to seedling germination and survival, small amounts of seed are particularly vulnerable to depredation by small mammals. Considerable flexibility is possible in timing site preparation to coincide with cone crops. Treatment can be apllied either before any logging takes place, between partial cuts, or after logging. In cut and leave strips, seedbed preparation can be carried out as a single operation, pre-scarifying the leave strips, post-scarifying the cut strips.

Broadcast burning is not recommended as a method of preparing sites for natural regeneration, as it rarely exposes enough mineral soil to be sufficiently receptive, and the charred organic surfaces are a poor seedbed for spruce. A charred surface may get too hot for good germination and may delay germination until fall, with subsequent overwinter mortality of unhardened seedlings. Piling and burning of logging slash, however, can leave suitable exposures of mineral soil.

Artificial Regeneration

With a view to reducing the time needed to produce planting stock, experiments were carried out with white spruce and three other coniferous species from Wisconsin seed in the longer, frost-free growing season in Florida, 125 vs. 265 days in central Wisconsin and northern Florida, respectively. As the species studied are adapted to long photoperiods, extended daylengths of 20 hours were applied in Florida. Other seedlings were grown under extended daylength in Wisconsin and with natural daylength in both areas. After two growing seasons, white spruce under long days in Florida were about the same as those in Wisconsin, but twice as tall as plants under natural Wisconsin photoperiods. Under natural days in Florida, with the short local photoperiod, white spruce was severely dwarfed and had a low rate of survival. Black spruce responded similarly. After two growing seasons, long day plants of all 4 species in Florida were well balanced, with good development of both roots and shoots, equaling or exceeding the minimum standards for 2+1 and 2+2 outplanting stock of Lake States species. Their survival when lifted in February and outplanted in Wisconsin equalled that of 2+2 Wisconsin-grown transplants. Artificial extension of the photoperiod in the northern Lake States greatly increased height increment of white and black spruces in the second growing season.

Optimum conditions for seedling growth have been determined for the production of containerized planting stock. Alternating day/night temperatures have been found more suitable than a constant temperature; at 400 lumens/m^2 light regime, a 28 °C/20 °C day/night temperatures have been recommended for white spruce. However, temperature optima are not necessarily the same at different ages and sizes. Tinus (1984) investigated the effects of combinations of day and night temperature on height, caliper, and dry weight of 4 seed sources of Engelmann spruce. The 4 seed sources appeared to have very similar temperature requirements, with night optima about the same of slightly lower than daylight optima.

Tree provenance is important in artificial regeneration. Good provenance takes into account suitable tree genetics and a good environmental fit for planted / seeded trees in a forest stand. The wrong genotype can lead to failed regeneration, or poor trees that are prone to pathogens and undesired outcomes.

Artificial regeneration has been a more common method involving planting because it is more dependable than natural regeneration. Planting can involve using seedlings (from a nursery), (un) rooted cuttings, or seeds.

Whichever method is chosen it can be assisted by tending techniques also known as intermediate stand treatments.

The fundamental genetic consideration in artificial regeneration is that seed and planting stock must be adapted to the planting environment. Most commonly, the method of managing seed and stock deployment is through a system of defined seed zones, within which seed and stock can be moved without risk of climatic maladaptation. Ontario adopted a seed zone system in the 1970s based on Hills' (1952) site regions and provincial resource district boundaries, but Ontario's seed zones are now based on homogeneous climatic regions developed with the Ontario Climate Model. The regulations stipulate that source-identified seedlots may be either a general collection, when only the seed zone of origin is known, or a stand collection from a specific latitude and longitude. The movement of general-collection seed and stock across seed zone boundaries is prohibited, but the use of stand-collection seed and stock in another seed zone is acceptable when the Ontario Climate Model shows that the planting site and place of seed origin are climatically similar. The 12 seed zones for white spruce in Quebec are based mainly on ecological regions, with a few modifications for administrative convenience.

Seed quality varies with source. Seed orchards produce seed of the highest quality, then, in order of decreasing seed quality produced, seed production areas and seed collection areas follow, with controlled general collections and uncontrolled general collections producing the least characterized seed.

Seeds

Dewinging, Extraction

When seed is first separated from cones it is mixed with foreign matter, often 2 to 5 times the volume of the seed. The more or less firmly attached membranous wings on the seed must be detached before it is cleaned of foreign matter. The testa must not incur damage during the dewinging process. Two methods have been used, dry and wet. Dry seed may be rubbed gently through a sieve that has a mesh through which only seed without wings can pass. Large quantities of seed can be processed in dewinging machines, which use cylinders of heavy wire mesh and rapidly revolving stiff brushes within to remove the wings. In the wet process, seed with wings attached are spread out 10 cm to 15 cm deep on a tight floor and slightly moistened throughout; light leather flails are used to free seed from the wings. Wang (1973) described a unique wet dewinging procedure using a cement mixer, used at the Petawawa tree seed processing facility. Wings of white and Norway spruce seed can be removed by dampening the seed slightly before it is run through a fanning mill for the last time. Any moistened seed must be dried before fermentation or moulding sets in.

Seed Viability

A fluorescein diacetate (FDA) biochemical viability test for several species of conifer seed, including white spruce, estimates the proportion of live seed (viability) in a seedlot, and hence the percentage germination of a seedlot. The accuracy of predicting percentage germination was within +/- 5 for most seedlots. White spruce seed can be tested for viability by an indirect method, such as the fluorescein diacetate (FDA) test or 'Ultra-sound'; or by the direct growth method of 'germination'. Samples of white spruce seed inspected by Toumey and Stevens (1928) varied in viability from 50% to 100%, but averaged 93%. Rafn (1915) reported 97% viability for white spruce seed.

Germinative Testing

The results of a germination test are commonly expressed as *germinative capacity* or a *germination percentage*, which is the percentage of seeds that germinate during a period of time, ending when germination is practically complete. During extraction and processing, white spruce seeds gradually lost moisture, and total germination increased. Mittal et al. (1987) reported that white spruce seed from Algonquin Park, Ontario, obtained the maximum rate (94% in 6 days) and 99% total germination in 21 days after 14-week pre-chilling. The pre-treatment of 1% sodium hypochlorite increased germinability.

Encouraged by Russian success in using ultrasonic waves to improve the germinative energy and percentage germination of seeds of agricultural crops, Timonin (1966) demonstrated benefits to white spruce germination after exposure of seeds to 1, 2, or 4 minutes of ultrasound generated by an M.S.E. ultrasonic disintegrator with a power consumption of 280 VA and power impact of 1.35 amperes. However, no seeds germinated after 6 minutes of exposure to ultrasound.

Seed Dormancy

Seed dormancy is a complex phenomenon and is not always consistent within species. Cold stratification of white spruce seed to break dormancy has been specified as a requirement, but Heit (1961) and Hellum (1968) regarded stratification as unnecessary. Cone handling and storage conditions affect dormancy in that cold, humid storage (5 °C, 75% to 95% relative humidity) of the cones prior to extraction seemingly eliminated dormancy by overcoming the need to stratify. Periods of cold, damp weather during the period of cone storage might provide natural cold (stratification) treatment. Once dormancy was removed in cone storage, subsequent kiln-drying and seed storage did not reactivate dormancy.

Haddon and Winston (1982) found a reduction in viability of stratified seeds after 2 years of storage and suggested that stress might have been caused by stratification, e.g., by changes in seed biochemistry, reduced embryo vigor, seed aging or actual damage to the embryo. They further questioned the quality of the 2-year-old seed even though high germination occurred in the samples that were not stratified.

Cold Stratification

Cold stratification is the term applied to the storing of seeds in (and, strictly, in layers with) a moist medium, often peat or sand, with a view to maintaining viability and overcoming dormancy. Cold stratification is the term applied to storage at near-freezing temperatures, even if no medium is used. A common method of cold stratification, is to soak seed in tap water for up to 24 h, superficially dry it, then store moist for some weeks or even months at temperatures just above freezing. Although Hellum (1968) found that cold stratification of an Alberta seed source led to irregular germination, with decreasing germination with increasing length of the stratification period, Hocking's (1972) paired test with stratified and nonstratified Alberta seed from several sources revealed no trends in response to stratification. Hocking suggested that seed maturity, handling, and storage needed to be controlled before the need for stratification could be determined. Later, Winston and Haddon (1981) found that the storage of white spruce cones for 4 weeks at 5 °C prior to extraction obviated the need for stratification.

Seed Ripeness

Seed maturity cannot be predicted accurately from cone flotation, cone moisture content, cone specific gravity; but the province of B.C. found embryo occupying 90% + of the corrosion cavity and megagametophyte being firm and whitish in colour are the best predictors for white spruce in B.C., and Quebec can forecast seed maturity some weeks in advance by monitoring seed development in relation to heat-sums and the phenological progression of the inflorescence of fireweed (*Epilobium angustifolium* L.), an associated plant species. Cone collection earlier than one week before seed maturity would reduce seed germination and viability during storage. Four stages of maturation were determined by monitoring carbohydrates, polyols, organic acids, respiration, and metabolic activity. White spruce seeds require a 6-week post-harvest ripening period in the cones to obtain maximum germinability, however, based on cumulative degree-days, seed from the same trees and stand showed 2-week cone storage was sufficient.

Plantation Establishment Criteria

Plantations may be considered successful when outplant performance satisfies certain criteria. The term "free growing" is applied in some jurisdictions. Ontario's "Free-to-Grow" (FTG) equivalent relates to a forest stand that meets a minimum stocking standard and height requirement, and is essentially free of competition from surrounding vegetation that might impede growth. The FTG concept was introduced with the advent of the Forest Management Agreement program in Ontario in 1980 and became applicable to all management units in 1986. Policy, procedures, and methodologies readily applicable by forest unit managers to assess the effectiveness of regeneration programs were still under development during the Class Environmental Assessment hearings.

In British Columbia, the Forest Practices Code (1995) governs performance criteria. To minimize the subjectivity of assessing deciduous competition as to whether or not a plantation is established, minimum specifications of number, health, height, and competition have been specified in British Columbia. However, minimum specifications are still subjectively set and may need to be fine-tuned in order to avoid unwarranted delay in according established status to a plantation. For example, a vigorous white spruce with a strong, multi-budded leading shoot and its crown fully exposed to light on 3 sides would not qualify as free-growing in the current British Columbia Code but would hardly warrant description as unestablished.

Competition

Competition arises when individual organisms are sufficiently close together to incur growth constraint through mutual modification of the local environment. Plants may compete for light, moisture and nutrients, but seldom for space *per se*. Vegetation management directs more of the site's resources into usable forest products, rather than just eliminating all competing plants. Ideally, site preparation ameliorates competition to levels that relieve the outplant of constraints severe enough to cause prolonged check.

The diversity of boreal and sub-boreal broadleaf-conifer mixed tree species stands, commonly referred to as the "mixedwoods", largely preclude the utility of generalizations and call for the development of management practices incorporating the greater inherent complexity of broadleaf-conifer mixtures, relative to single-species or mixed-species conifer forest. After harvesting or other disturbance, mixedwood stands commonly enter a prolonged period in which hardwoods overtop the coniferous component, subjecting them to intense competition in an understorey. It is well established that the regeneration and growth potential of understorey conifers in mixedwood stands is correlated to the density of competing hardwoods. To help apply "free-to-grow" regulations in British Columbia and Alberta, management guidelines based on distance-dependent relations within a limited radius of crop trees were developed, but Lieffers et al. (2002) found that free-growing stocking standards did not adequately characterize light competition between broadleaf and conifer components in boreal mixedwood stands, and further noted that adequate sampling using current approaches would be operationally prohibitive.

Many promising plantations have failed through lack of tending. Young crop trees are often ill-equipped to fight it out with competition resurgent following initial site preparation and planting.

Perhaps the most direct evaluation of the effect of competition on plantation establishment is provided by an effective herbicide treatment. The fact that herbicide treatment does not always produce positive results should not obscure the demonstrated potential of herbicides for significantly promoting plantation establishment. Factors that can vitiate the effectiveness of a herbicide treatment include: weather, especially temperature, prior to and during application; weather, especially wind, during application; weather, especially precipitation, in the 12 to 24 hours after application; vegetation characteristics, including species, size, shape, phenological stage, vigour, and distribution of weeds; crop characteristics, including species, phenology, and condition; the effects of other treatments, such as preliminary shearblading, burning or other prescribed or accidental site preparation; and the herbicide used, including dosage, formulation, carrier, spreader, and mode of application. There is a lot that can go wrong, but a herbicide treatment can be as good or better than any other method of site preparation.

Competition Indices

The study of competition dynamics requires both a measure of the competition level and a measure of crop response. Various competition indices have been developed, e.g., by Bella (1971) and Hegyi (1974) based on stem diameter, by Arney (1972), Ek and Monserud (1974), and Howard and Newton (1984) based on canopy development, and Daniels (1976), Wagner (1982), and Weiner (1984) with proximity-based models. Studies generally considered tree response to competition in terms of absolute height or basal area, but Zedaker (1982) and Brand (1986) sought to quantify crop tree size and environmental influences by using relative growth measures.

Tending

Tending is the term applied to pre-harvest silvicultural treatment of forest crop trees at any stage after initial planting or seeding. The treatment can be of the crop itself (e.g., spacing, pruning, thinning, and improvement cutting) or of competing vegetation (e.g., weeding, cleaning).

Planting

How many trees per unit area (spacing) that should be planted is not an easily answered question. Establishment density targets or regeneration standards have commonly been based on traditional practice, with the implicit aim of getting the stand quickly to the free-to-grow stage. Money is wasted if more trees are planted than are needed to achieve desired stocking rates, and the chance to establish other plantations is proportionately diminished. Ingress (natural regeneration) on a site is difficult to predict and often becomes surprisingly evident only some years after planting has been carried out. Early stand development after harvesting or other disturbance undoubtedly varies greatly among sites, each of which has its own peculiar characteristics.

For all practical purposes, the total volume produced by a stand on a given site is constant and optimum for a wide range of density or stocking. It can be decreased, but not increased, by altering the amount of growing stock to levels outside this range. Initial density affects stand development in that close spacing leads to full site utilization more quickly than wider spacing. Economic operability can be advanced by wide spacing even if total production is less than in closely spaced stands.

Beyond the establishment stage, the relationship of average tree size and stand density is very important. Various density-management diagrams conceptualizing the density-driven stand dynamics have been developed. Smith and Brand's (1988) diagram has mean tree volume on the vertical axis and the number of trees/ha on the horizontal axis: a stand can either have many little trees or a few big ones. The self-thinning line shows the largest number of trees of a given size/ha that can be carried at any given time. However, Willcocks and Bell (1995) caution against using such diagrams unless specific knowledge of the stand trajectory is known.

In the Lake States, plantations have been made with the spacing between trees varying from 3 by 3 to 10 by 10 feet (0.9 m by 0.9 m to 3.0 m by 3.0 m). Kittredge recommended that no fewer than 600 established trees per acre (1483/ha) be present during the early life of a plantation. To insure this, at least 800 trees per acre (1077/ha) should be planted where 85% survival may be expected, and at least 1200/ac (2970/ha) if only half of them can be expected to live. This translates into recommended spacings of 5 by 5 to 8 by 8 feet (1.5 m by 1.5 m to 2.4 m by 2.4 m) for plantings of conifers, including white spruce in the Lake States.

Enrichment Planting

A strategy for enhancing natural forests' economic value is to increase their concentration of economically important, indigenous tree species by planting seeds or seedlings for future harvest, which can be accomplished with enrichment planting (EP). This means increasing the planting density (i.e., the numbers of plants per hectare) in an already growing forest stand."

Release Treatments

- Weeding: A process of getting rid of saplings' or seedlings' competition by mowing, application of herbicide, or other method of removal from the surroundings.
- Cleaning: Release of select saplings from competition by overtopping trees of a comparable age. The treatment favors trees of a desired species and stem quality.
- Liberation cutting: A treatment that releases tree seedling or saplings by removing older overtopping trees.

Spacing

Over-crowded regeneration tends to stagnate. The problem is aggravated in species that have little self-pruning ability, such as white spruce. Spacing is a thinning (of natural regeneration), in which all trees other than those selected for retention at fixed intervals are cut. The term juvenile spacing is used when most or all of the cut trees are unmerchantable. Spacing can be used to obtain any of a wide range of forest management objectives, but it is especially undertaken to reduce density and control stocking in young stands and prevent stagnation, and to shorten the rotation, i.e., to speed the production of trees of a given size. Volume growth of individual trees and the merchantable growth of stands are increased. The primary rationale for spacing is that thinning is the projected decline in maximum allowable cut. And since wood will be concentrated on fewer, larger, and more uniform stems, operating and milling costs will be minimized.

Methods for spacing may be: manual, using various tools, including power saws, brush saws, and clippers; mechanical, using choppersand mulchers; chemical; or combinations of several methods. One treatment has had notable success in spacing massively overstocked (<100 000 stems/ha) natural regeneration of spruce and fir in Maine. Fitted to helicopter, the Thru-Valve boom emits herbicide spray droplets 1000 µm to 2000 µm in diameter at very low pressure. Swaths 1.2 m wide and leave strips 2.4 m wide were obtained with "knife-edge" precision when the herbicide was applied by helicopter flying at a height of 21 m at a speed of 40–48 km/h. It seems likely that no other method could be as cost-effective.

Twenty years after spacing to 2.5 × 2.5 m, 30-year-old mixed stands of balsam fir and white spruce in the Green River watershed, New Brunswick, averaged 156.9 m^3/ha.

A spacing study of 3 conifers (white spruce, red pine and jack pine) was established at Moodie, Manitoba, on flat, sandy, nutritionally poor soils with a fresh moisture regime. Twenty years after planting, red pine had the largest average dbh, 15% greater than jack pine, while white spruce dbh was less than half that of the pines. Crown width showed a gradual increase with spacing for all 3 conifers. Results to date were suggesting optimum spacings between 1.8 m and 2.4 m for both pines; white spruce was not recommended for planting on such sites.

Comparable data are generated by espacement trials, in which trees are planted at a range of densities. Spacings of 1.25 m, 1.50 m, 1.75 m, 2.00 m, 2.50 m, and 3.00 m on 4 site classes were used in the 1922 trial at Petawawa, Ontario. In the first of 34 old field white spruce plantations used to investigate stand development in relation to spacing at Petawawa, Ontario, regular rows were planted at average spacings of from 4 × 4 to 7 × 7 feet (1.22 m × 1.22 m to 2.13 m × 2.13 m). Spacings up to 10 × 10 feet (3.05 m × 3.03 m) were subsequently included in the study. Yield tables based on 50 years of data showed:

a) Except for merchantable volumes at age 20 and site classes 50 and 60, closer spacings gave greater standing volumes at all ages than did wider spacings, the relative difference decreasing with age.

b) Merchantable volume as a proportion of total volume increases with age, and is greater at wider than at closer spacings.

c) Current annual volume increment culminates sooner at closer than at wider spacings.

A smaller espacement trial, begun in 1951 near Thunder Bay, Ontario, included white spruce at spacings of 1.8 m, 2.7 m, and 3.6 m. At the closest spacing, mortality had begun at 37 years, but not at the wider spacings.

The oldest interior spruce espacement trial in British Columbia was established in 1959 near Houston in the Prince Rupert Forest Region. Spacings of 1.2 m, 2.7 m, 3.7 m, and 4.9 m were used, and trees were measured 6, 12, 16, 26, and 30 years after planting. At wide espacements, trees developed larger diameters, crowns, and branches, but (at 30 years) basal area and total volume/ha were greatest in the closest espacement (Table 6.38). In more recent trials in the Prince George Region of British Columbia (Table 6.39) and in Manitoba, planting density of white spruce had no effect on growth after up to 16 growing seasons, even at spacings as low as 1.2 m. The slowness of juvenile growth and of crown closure delay the response to intra-competition. Initially, close spacing might even provide a positive nurse effect to offset any negative response to competition.

Thinning

Thinning is an operation that artificially reduces the number of trees growing in a stand with the aim of hastening the development of the remainder. The goal of thinning is to control the amount and distribution of available growing space. By altering stand density, foresters can influence the growth, quality, and health of residual trees. It also provides an opportunity to capture mortality and cull the commercially less desirable, usually smaller and malformed, trees. Unlike regeneration treatments, thinnings are not intended to establish a new tree crop or create permanent canopy openings.

Thinning greatly influences the ecology and micro-meteorology of the stand, lowering the inter-tree competition for water. The removal of any tree from a stand has repercussions on the remaining trees both above-ground and below. Silvicultural thinning is a powerful tool that can be used to influence stand development, stand stability, and the characteristics of the harvestable products.

When considering intensive conifer plantations designed for maximum production, it is essential to remember that tending and thinning regimes and wind and snow damage are intimately related.

Previous studies have demonstrated that repeated thinnings over the course of a forest rotation increase carbon stores relative to stands that are clear-cut on short rotations and that the carbon benefits differ according to thinning method (e.g., thinning from above versus below).

Precommercial Thinning

When natural regeneration or artificial seeding has resulted in dense, overstocked young stands, natural thinning will in most cases eventually reduce stocking to more silviculturally desirable levels. But by the time some trees reach merchantable size, others will be overmature and defective, and others will still be unmerchantable. The yield of merchantable wood can be greatly increased and the rotation shortened by precommercial thinning. Mechanical and chemical methods have been applied, but their costliness has militated against their ready adoption.

Pruning

Pruning, as a silvicultural practice, refers to the removal of the lower branches of the young trees (also giving the shape to the tree) so clear knot-free wood can subsequently grow over the branch stubs. Clear knot-free lumber has a higher value. Pruning has been extensively carried out in the Radiata pine plantations of New Zealand and Chile, however the development of Finger joint technology in the production of lumber and mouldings has led to many forestry companies reconsidering their pruning practices. "Brashing" is an alternative name for the same process. Pruning can be done to all trees, or more cost effectively to a limited number of trees. There are two types of pruning: natural or self-pruning and artificial pruning. Most cases of self-pruning happen when branches do not receive enough sunlight and die. Wind can also take part in natural pruning which can break branches. Artificial pruning is where people are paid to come and cut the branches. Or it can be natural, where trees are planted close enough that the effect is to cause self-pruning of low branches as energy is put into growing up for light reasons and not branchiness.

Stand Conversion

The term *stand conversion* refers to a change from one silvicultural system to another and includes *species conversion*, i.e., a change from one species (or set of species) to another. Such change can be effected intentionally by various silvicultural means, or incidentally by default e.g., when high-grading has removed the coniferous content from a mixedwood stand, which then becomes exclusively self-perpetuating aspen. In general, such sites as these are the most likely to be considered for conversion.

Growth and Yield

In discussing yields that might be expected from the Canadian spruce forests, Haddock (1961) noted that Wright's (1959) quotation of spruce yields in the British Isles of 220 cubic feet per acre (15.4 m^3/ha) per year and in Germany of 175 cubic feet per acre (12.25 m^3/ha) per year was misleading, at least if it was meant to imply that such yields might be approached in the Boreal Forest Region of Canada. Haddock thought that Wright's suggestion of 20 to 40 (average 30) cubic feet per acre (1.4 m^3/ha to 2.8 m^3/ha (average 2.1 m^3/ha) per year was more reasonable, but still somewhat optimistic.

The principal way forest resource managers influence growth and yield is to manipulate the mixture of species and number (density) and distribution (stocking) of individuals that form the canopy of the stand. Species composition of much of the boreal forest in North America already differs greatly from its pre-exploitation state. There is less spruce and more hardwoods in the second-growth forest than in the original forest; Hearnden et al. (1996) calculated that the spruce cover type had declined from 18% to only 4% of the total forested area in Ontario. Mixedwood occupies a greater proportion of Ontario's second-growth forest (41%) than in the original (36%), but its component of white spruce is certainly much diminished.

Growth performance is certainly influenced by site conditions and thus by the kind and degree of site preparation in relation to the nature of the site. It is important to avoid the assumption that site preparation of a particular designation will have a particular silvicultural outcome. Scarification, for instance, not only covers a wide range of operations that scarify, but also any given way of scarifying can have significantly different results depending on site conditions at the time of treatment. In point of fact, the term is commonly misapplied. *Scarification* is defined as "Loosening the top soil of open areas, or breaking up the forest floor, in preparation for regenerating by direct seeding or natural seedfall", but the term is often misapplied to practices that include scalping, screefing, and blading, which pare off low and surface vegetation, together with most off its roots to expose a weed-free surface, generally in preparation for sowing or planting thereon.

Thus, it is not surprising that literature can be used to support the view that the growth of seedlings on scarified sites is much superior to that of growth on similar sites that have not been scarified, while other evidence supports the contrary view that scarification can reduce growth. Detrimental results can be expected from scarification that impoverishes the rooting zone or exacerbates edaphic or climatic constraints.

Burning site preparation has enhanced spruce seedling growth, but it must be supposed that burning could be detrimental if the nutrient capital is significantly depleted.

An obvious factor greatly influencing regeneration is competition from other vegetation. In a pure stand of Norway spruce, for instance, Roussel (1948) found the following relationships:

Percent cover (%)	Vegetation Description
Below 1	No vegetation
1-3	Moss carpet with a few fir seedlings
4-10	Herbaceous plants appear
10-25	Bramble, herbs, fairly vigorous spruce seedlings
>25	Herbs, brambles very dense, vigorous, no moss

A factor of some importance in solar radiation–reproduction relationships is excess heating of the soil surface by radiation. This is especially important for seedlings, such as spruce, whose first leaves do not shade the base of the stem at the soil surface. Surface temperatures in sandy soils on occasion reach lethal temperatures of 50 °C to 60 °C.

Common Methods of Harvesting

Silvicultural regeneration methods combine both the harvest of the timber on the stand and re-establishment of the forest. The proper practice of sustainable forestry should mitigate the potential negative impacts, but all harvest methods will have some impacts on the land and residual stand. The practice of sustainable forestry limits the impacts such that the values of the forest are maintained in perpetuity. Silvicultural prescriptions are specific solutions to a specific set of circumstances and management objectives. Following are some common methods:

Clearcut Harvesting

Conventional clearcut harvesting is relatively simple: all trees on a cutblock are felled and bunched with bunches aligned to the skidding direction, and a skidder then drags the bunches to the closest log deck. Feller-buncher operators concentrate on the width of the felled swath, the number of trees in a bunch, and the alignment of the bunch. Providing a perimeter boundary is felled during daylight, night-shift operations can continue without the danger of trespassing beyond the block. Productivity of equipment is maximized because units can work independently of one another.

Clearcutting

An even-aged regeneration method that can employ either natural or artificial regeneration. It involves the complete removal of the forest stand at one time. Clearcutting can be biologically appropriate with species that typically regenerate from stand replacing fires or other major disturbances, such as Lodgepole Pine (*Pinus contorta*). Alternatively, clearcutting can change the dominating species on a stand with the introduction of non-native and invasive species as was shown at the Blodgett Experimental Forest near Georgetown California. Additionally, clearcutting can prolong slash decomposition, expose soil to erosion, impact visual appeal of a landscape and remove essential wildlife habitat. It is particularly useful in regeneration of tree species such as Douglas-fir (*Pseudotsuga menziesii*) which is shade intolerant.. In addition, the general public's distaste for even-aged silviculture, particularly clearcutting, is likely to result in a greater role for uneven-aged management on public lands as well. Across Europe, and in parts of North America, even-aged, production-orientated and intensively managed plantations are beginning to be regarded in the same way as old industrial complexes: something to abolish or convert to something else.

Clearcutting will impact many site factors important in their effect on regeneration, including air and soil temperatures. Kubin and Kemppainen (1991), for instance, measured temperatures in northern Finland from 1974 through 1985 in 3 clear-felled areas and in 3 neighouring forest stands dominated by Norway spruce. Clear felling had no significant influence on air temperature at 2 m above the ground surface, but the daily air temperature maxima at 10 cm were greater in the clear-felled area than in the uncut forest, while the daily minima at 10 cm were lower. Night frosts were more common in the clear-felled area. Daily soil temperatures at 5 cm depth were 2 °C to 3 °C greater in the clear-felled area than in the uncut forest, and temperatures at depths of 50 cm and 100 cm were 3 °C to 5 °C greater. The differences between the clear-felled and uncut areas did not diminish during the 12 years following cutting.

Coppicing

A regeneration method which depends on the sprouting of cut trees. Most hardwoods, the coast redwood, and certain pines naturally sprout from stumps and can be managed through coppicing. Coppicing is generally used to produce fuelwood, pulpwood, and other products dependent on small trees. A close relative of coppicing is pollarding. Three systems of coppice woodland management are generally recognized: simple coppice, coppice with standards, and the coppice selection system.

- In Compound coppicing or coppicing with standards, some of the highest quality trees are retained for multiple rotations in order to obtain larger trees for different purposes.

Direct Seeding

Prochnau (1963), 4 years after sowing, found that 14% of viable white spruce seed sown on mineral soil had produced surviving seedlings, at a seed:seedling ratio of 7.1:1. With Engelmann spruce, Smith and Clark (1960) obtained average 7th year seed:seedling ratios of 21:1 on scarified seedbeds on dry sites, 38:1 on moist sites, and 111:1 on litter seedbeds.

Group Selection

The group selection method is an uneven-aged regeneration method that can be used when mid-tolerant species regeneration is desired. The group selection method can still result in residual stand damage in dense stands, however directional falling can minimize the damage. Additionally, foresters can select across the range of diameter classes in the stand and maintain a mosaic of age and diameter classes.

Méthode Du Contrôle

Classical European silviculture achieved impressive results with systems such as Henri Biolley's *méthode du contrôle* in Switzerland, in which the number and size of trees harvested were determined by reference to data collected from every tree in every stand measured every 7 years.

While not designed to be applied to boreal mixedwoods, the *méthode du contrôle* is described briefly here to illustrate the degree of sophistication applied by some European foresters to the management of their forests. Development of management techniques that allowed for stand development to be monitored and guided into sustainable paths were in part a response to past experience, particularly in Central European countries, of the negative effects of pure, uniform stands with species often unsuited to the site, which greatly increased the risk of soil degradation and biotic diseases. Increased mortality and decreased increment generated widespread concern, especially after reinforcement by other environmental stresses.

More or less uneven-aged, mixed forests of preponderantly native species, on the other hand, treated along natural lines, have proved to be healthier and more resistant to all kinds of external dangers; and in the long run such stands are more productive and easier to protect.

However, irregular stands of this type are definitely more difficult to manage—new methods and techniques had to be sought particularly for the establishment of inventories, as well as increment control and yield regulation. In Germany, for instance, since the beginning of the nineteenth-century under the influence of G.L. Hartig (1764–1837), yield regulation has been effected almost exclusively by allotment or formula methods based on the conception of the uniform normal forest with a regular succession of cutting areas.

In France, on the other hand, efforts were made to apply another kind of forest management, one that aimed to bring all parts of the forest to a state of highest productive capacity in perpetuity. In 1878, the French forester A. Gurnaud (1825–1898) published a description of a *méthode du contrôle* for determining increment and yield. The method was based on the fact that through careful, selective harvesting, the productivity of the residual stand can be improved, because timber is removed as a cultural operation. In this method, the increment of stands is accurately determined periodically with the object of gradually converting the forest, through selective management and continuous experimentation, to a condition of equilibrium at maximum productive capacity.

Henri Biolley (1858–1939) was the first to apply Gurnaud's inspired ideas to practical forestry. From 1890 on, he managed the forests of his Swiss district according to these principles, devoting himself for almost 50 years to the study of increment and a treatment of stands directed towards the highest production, and proving the practicability of the check method. In 1920, he published this study giving a theoretical basis of management of forests under the check method, describing the procedures to be applied in practice (which he partly developed and simplified), and evaluating the results.

Biolley's pioneering work formed the basis upon which most Swiss forest management practices were later developed, and his ideas have been generally accepted. Today, with the trend of intensifying forest management and productivity in most countries, the ideas and application of careful, continuous treatment of stands with the aid of the volume check method are meeting with ever-growing interest. In Britain and Ireland, for example, there is increased application of Continuous Cover Forestry principles to create permanently irregular structures in many woodlands.

Patch Cut

Row and Broadcast Seeding

Spot and row seeders use less seed that does broadcast ground or aerial seeding but may induce clumping. Row and spot seeding confer greater ability to control seed placement than does broadcast seeding. Also, only a small percentage of the total area needs to be treated.

In the aspen type of the Great Lakes region, direct sowing of the seed of conifers has usually failed. However, Gardner (1980) after trials in Yukon, which included broadcast seeding of white spruce seed at 2.24 kg/ha that secured 66.5% stocking in the Scarified Spring Broadcast treatment 3 years after seeding, concluded that the technique held “considerable promise”.

Seed-Tree

An even-aged regeneration method that retains widely spaced residual trees in order to provide uniform seed dispersal across a harvested area. In the seed-tree method, 2-12 seed trees per acre (5-30/ha) are left standing in order to regenerate the forest. They will be retained until regeneration has become established at which point they may be removed. It may not always be economically viable or biologically desirable to re-enter the stand to remove the remaining seed trees. Seed-tree cuts can also be viewed as a clearcut with natural regeneration and can also have all of the problems associated with clearcutting. This method is most suited for light-seeded species and those not prone to windthrow.

Selection Systems

Selection systems are appropriate where uneven stand structure is desired, particularly where the need to retain continuous cover forest for aesthetic or environmental reasons outweighs other management considerations. Selection logging has been suggested as being of greater utility than shelterwood systems in regenerating old-growth Engelmann Spruce Sub-alpine Fir (ESSF) stands in southern British Columbia. In most areas, selection logging favours regeneration of fir more than the more light-demanding spruce. In some areas, selection logging can be expected to favour spruce over less tolerant hardwood species (Zasada 1972) or lodgepole pine.

Shelter Spot Seeding

The use of shelters to improve germination and survival in spot seedings seeks to capture the benefits of greenhouse culture, albeit miniature. The Hakmet seed shelter, for instance, is a semi-transparent plastic cone 8 cm high, with openings of 7 cm diameter in the 7.5 cm diameter base and 17 mm diameter in the 24 mm diameter top. This miniature greenhouse increases air humidity, reduces soil desiccation, and raises air and soil temperatures to levels more favourable to germination and seedling growth than those offered by unprotected conditions. The shelter is designed to break down after a few years of exposure to ultraviolet radiation.

Seed shelters and spring sowing significantly improved stocking compared with bare spot seeding, but sheltering did not significantly improve growth. Stocking of bare seedspots was extremely low, possibly due to smothering of seedlings by abundant broadleaf and herbaceous litter, particularly that from aspen and red raspberry, and exacerbated by strong competition from graminoids and raspberry.

Cone shelters (Cerkon™) usually produced greater survival than unsheltered seeding on scarified seedspots in trials of direct seeding techniques in interior Alaska, and funnel shelters (Cerbel™) usually produced greater survival than unsheltered seeding on non-scarified seedspots. Both shelter types are manufactured by AB Cerbo in Trollhättan, Sweden. Both are made of light-degradable, white, opaque plastic, and are 8 cm high when installed.

White spruce seed was sown in Alaska on a burned site in summer 1984, and protected by white plastic cones on small spots scarified by hand, or by white funnels placed directly into the residual ash and organic material. A group of 6 ravens (*Corvus corax*) was observed in the area about 1 week after sowing was completed in mid-June. Damage averaged 68% with cones and 50% with funnels on an upland area, and 26% with funnels on a floodplain area. Damage by ravens was only 0.13% on unburned but otherwise similar areas.

In seeding trials in Manitoba between 1960 and 1966 aimed at converting aspen stands to spruce–aspen mixedwoods, 1961 scarification in the Duck Mountain Provincial Forest remained receptive to natural seeding for many years.

Shelterwood

In general terms, the shelterwood system is a series of partial cuts that removes the trees of an existing stand over several years and eventually culminates in a final cut that creates a new even-aged stand. It is an even-aged regeneration method that removes trees in a series of three harvests: 1) Preparatory cut; 2) Establishment cut; and 3) Removal cut. The success of practising a shelterwood system is closely related to: 1. the length of the regeneration period, i.e. the time from the shelterwood cutting to the date when a new generation of trees has been established; 2.the quality of the new tree stand with respect to stand density and growth; and 3.the value increment of the shelter trees. Information on the establishment, survival and growth of seedlings influenced by the cover of shelter trees, as well as on the growth of these trees, is needed as a basis for modelling the economic return of practising a shelterwood system. The method's objective is to establish new forest reproduction under the shelter of the retained trees. Unlike the seed-tree method, residual trees alter understory environmental conditions (i.e. sunlight, temperature, and moisture) that influence tree seedling growth. This method can also find a middle ground with the light ambiance by having less light accessible to competitors while still being able to provide enough light for tree regeneration. Hence, shelterwood methods are most often chosen for site types characterized by extreme conditions, in order to create a new tree generation within a reasonable time period. These conditions are valid foremost on level ground sites which are either dry and poor or moist and fertile.

Shelterwood Systems

Shelterwood systems involve 2, 3, or exceptionally more partial cuttings. A final cut is made once adequate natural regeneration has been obtained. The shelterwood system is most commonly applied as a 2-cut uniform shelterwood, first an initial regeneration (seed) cut, the second a final harvest cut. In stands less than 100 years old, a light preparatory cut can be useful. A series of intermediate cuts at intervals of 10–20 years has been recommended for intensively managed stands.

From operational or economic standpoints, however, there are disadvantages to the shelterwood system: harvesting costs are higher; trees left for deferred cutting may be damaged during the regeneration cut or related extraction operations; the increased risk of blowdown threatens the seed source; damage from bark beetles is likely to increase; regeneration may be damaged during the final cut and related extraction operations; the difficulty of any site preparation would be increased; and incidental damage to regeneration might be caused by any site preparation operations.

Single-Tree Selection

The single-tree selection method is an uneven-aged regeneration method most suitable when shade tolerant species regeneration is desired. It is typical for older and diseased trees to be removed, thus thinning the stand and allowing for younger, healthy trees to grow. Single-tree selection can be very difficult to implement in dense or sensitive stands and residual stand damage can occur. This method is also disturbs the canopy layer the least out of all other methods.

Spot Seeding

Spot seeding was found to be the most economical and reliable of the direct seeding methods for converting aspen and paper birch to spruce and pine. In the Chippewa National Forest (Lake States), seed-spot sowing of 10 seeds each of white spruce and white pine under 40-year aspen after different degrees of cutting on gave second-season results clearly indicating the need to remove or disturb the forest floor to obtain germination of seeded white spruce and white pine.

Spot seeding of coniferous seed, including white spruce, has had occasional success, but several constraining factors commonly limit germination success: the drying out of the forest floor before the roots of germinants reach underlying moisture reserves; and, particularly under hardwoods, the smothering of small seedlings by snow-pressed leaf litter and lesser vegetation. Kittredge and Gervorkiantz (1929) determined that removal of the aspen forest floor increased germination percentage after the second season in seed spots of both white pine and white spruce, in 4 plots, from 2.5% to 5%, from 8% to 22%, from 1% to 9.5%, and from 0% to 15%.

Spot seeding requires less seed than broadcast seeding and tends to achieve more uniform spacing, albeit sometimes with clumping. The devices used in Ontario for manual spot seeding are the "oil can" seeder, seeding sticks, and shakers. The oil can is a container fitted with a long spout through which a predetermined number of seeds are released with each flick of the seeder.

Strip Cutting

Harvesting cutblocks where only a portion of the trees are to be removed is very different from clearcutting. First, trails must be located to provide access for the felling and skidding/forwarding equipment. These trails must be carefully located to ensure that the trees remaining meet the desired quality criteria and stocking density. Second, the equipment must not damage the residual stand. The further desiderata are outlined by Sauder (1995).

The dearth of seed and a deficiency of receptive seedbeds were recognized as major reasons for the lack of success of clearcut harvesting. One remedy attempted in British Columbia and Alberta has been alternate strip cutting. The greater seed source from uncut trees between the cut strips, and the disturbance to the forest floor within the cut strips could be expected to increase the amount of natural regeneration. Trees were cut to a diameter limit in the cut strips, but large trees in the leave strips often proved too much of a temptation and were cut too, thus removing those trees that would otherwise have been the major source of seed.

An unfortunate consequence of strip thinning was the build-up of spruce beetle populations. Shaded slash from the initial cut, together with an increase in the number of windthrown trees in the leave strips, provided conditions ideally suited to the beetle.

Underplanting

DeLong et al. (1991) suggested underplanting 30- to 40-year-old aspen stands, on the basis of the success of natural spruce in regenerating under stands of such stands: "By planting, spacing can be controlled enabling easier protection of the spruce during stand entry for harvesting of the aspen overstorey".

Variable Retention

A harvesting and regeneration method which is a relatively new silvicultural system that retains forest structural elements (stumps, logs, snags, trees, understory spieces and undisturbed layers of forest floor) for at least one rotation in order to preserve environmental values associated with structurally complex forests.

"Uneven-aged and even-aged methods differ in the scale and intensity of disturbance. Uneven-aged methods maintain a mix of tree sizes or ages within a habitat patch by periodically harvesting individual or small groups of trees, Even-aged methods harvest most or all of the overstory and create a fairly uniform habitat patch dominated by trees of the same age". Even-aged management systems have been the prime methods to use when studying the effects on birds.

Mortality

A survey in 1955–56 to determine survival, development, and the reasons for success or failure of conifer pulpwood plantations (mainly of white spruce) in Ontario and Quebec up to 32 years old found that the bulk of the mortality occurred within the first 4 years of planting, unfavourable site and climate being the main causes of failure.

Advance Growth

Naturally regenerated trees in an understorey prior to harvesting constitute a classic case of good news and bad news. Understorey white spruce is of particular importance in mixedwoods dominated by aspen, as in the B15, B18a, and B19a Sections of Manitoba, and elsewhere. Until the latter part of the last century, white spruce understorey was mostly viewed as money in the bank on a long-term, low interest deposit, with final yield to be realized after slow natural succession, but the resource became increasingly threatened with the intensification of harvesting of aspen. White spruce plantations on mixedwood sites proved expensive, risky, and generally unsuccessful. This prompted efforts to see what might be done about growing aspen and white spruce on the same landbase by protecting existing white spruce advance growth, leaving a range of viable crop trees during the first cut, then harvesting both hardwoods and spruce in the final cut. Information about the understorey component is critical to spruce management planning. The ability of then current harvesting technology and crews employed to provide adequate protection for white spruce understories was questioned by Brace and Bella. Specialized equipment and training, perhaps with financial incentives, may be needed to develop procedures that would confer the degree of protection needed for the system to be feasible. Effective understorey management planning requires more than improved mixedwood inventory.

Avoidance of damage to the understorey will always be a desideratum. Sauder's (1990) paper on mixedwood harvesting describes studies designed to evaluate methods of reducing non-trivial damage to understorey residuals that would compromise their chance of becoming a future crop tree. Sauder concluded that: (1) operational measures that protected residual stems may not unduly increase costs, (2) all felling, conifers and hardwoods, needs to be done in one operation to minimize the entry of the feller-buncher into the residual stand, (3) several operational procedures can reduce understorey damage, some of them without incurring extra costs, and (4) successful harvesting of treatment blocks depends primarily on the intelligent location of skid trails and landings. In summary, the key to protecting the white spruce understorey without sacrificing logging efficiency is a combination of good planning, good supervision, the use of appropriate equipment, and having conscientious, well-trained operators.Even the best plan will not reduce understorey damage unless its implementation is supervised.

New stands need to be established to provide for future supply of commercial white spruce from 150 000 ha of boreal mixedwoods in 4 of Rowe's (1972) regional Forest Sections straddling Alberta, Saskatchewan, and Manitoba, roughly from Peace River AB to Brandon MB. In the 1980s, with harvesting using conventional equipment and procedures, a dramatic increase in the demand for aspen posed a serious problem for the associated spruce understorey. Formerly, white spruce in the understories had developed to commercial size through natural succession under the protection of the hardwoods. Brace articulated a widespread concern: "The need for protection of spruce as a component of boreal mixedwoods goes beyond concern for the future commercial softwood timber supply. Concerns also include fisheries and wildlife habitat, aesthetics and recreation, a general dissatisfaction with cleacutting in mixedwoods and a strong interest in mixedwood perpetuation, as expressed recently in 41 public meetings on forestry development in northern Alberta...".

On the basis of tests of 3 logging systems in Alberta, Brace (1990) affirmed that significant amounts of understorey can be retained using any of those systems provided that sufficient effort is directed towards protection. Potential benefits would include increased short-term softwood timber supply, improved wildlife habitat and cutblock aesthetics, as well as reduced public criticism of previous logging practices. Stewart et al. (2001) developed statistical models to predict the natural establishment and height growth of understorey white spruce in the boreal mixedwood forest in Alberta using data from 148 permanent sample plots and supplementary information about height growth of white spruce regeneration and the amount and type of available substrate. A discriminant model correctly classified 73% of the sites as to presence or absence of a white spruce understorey, based on the amount of spruce basal area, rotten wood, ecological nutrient regime, soil clay fraction, and elevation, although it explained only 30% of the variation in the data. On sites with a white spruce understorey, a regression model related the abundance of regeneration to rotten wood cover, spruce basal area, pine basal area, soil clay fraction, and grass cover ($R^2 = 0.36$). About half the seedlings surveyed grew on rotten wood, and only 3% on mineral soil, and seedlings were 10 times more likely to have established on these substrates than on litter. Exposed mineral soil covered only 0.3% of the observed transect area.

Advance Growth Management

Advance growth management, i.e., the use of suppressed understorey trees, can reduce reforestation costs, shorten rotations, avoid denuding the site of trees, and also reduce adverse impacts on aesthetic, wildlife, and watershed values. To be of value, advance growth must have acceptable species composition and distribution, have potential for growth following release, and not be vulnerable to excessive damage from logging.

The age of advance growth is difficult to estimate from its size, as white that appears to be 2- to 3-year-old may well be more than 20 years old. However, age does not seem to determine the ability of advance growth of spruce to respond to release, and trees older than 100 years have shown rapid rates of growth after release. Nor is there a clear relationship between the size of advance growth and its growth rate when released.

Where advance growth consists of both spruce and fir, the latter is apt to respond to release more quickly than the former, whereas spruce does respond. If the ratio of fir to spruce is large, however, the greater responsiveness to release of fir may subject the spruce to competition severe enough to negate much of the effect of release treatment. Even temporary relief from shrub competition has increased height growth rates of white spruce in northwestern New Brunswick, enabling the spruce to overtop the shrubs.

Site Preparation

Site preparation is any of various treatments applied to a site in order to ready it for seeding or planting. The purpose is to facilitate the regeneration of that site by the chosen method. Site preparation may be designed to achieve, singly or in any combination: improved access, by reducing or rearranging slash, and amelioration of adverse forest floor, soil, vegetation, or other biotic factors. Site preparation is undertaken to ameliorate one or more constraints that would otherwise be likely to thwart the objectives of management. A valuable bibliography on the effects of soil temperature and site preparation on subalpine and boreal tree species has been prepared by McKinnon et al. (2002).

Site preparation is the work that is done before a forest area is regenerated. Some types of site preparation are burning.

Burning

Broadcast burning is commonly used to prepare clearcut sites for planting, e.g., in central British Columbia, and in the temperate region of North America generally.

Prescribed burning is carried out primarily for slash hazard reduction and to improve site conditions for regeneration; all or some of the following benefits may accrue:

a) Reduction of logging slash, plant competition, and humus prior to direct seeding, planting, scarifying or in anticipation of natural seeding in partially cut stands or in connection with seed-tree systems.

b) Reduction or elimination of unwanted forest cover prior to planting or seeding, or prior to preliminary scarification thereto.

c) Reduction of humus on cold, moist sites to favour regeneration.

d) Reduction or elimination of slash, grass, or brush fuels from strategic areas around forested land to reduce the chances of damage by wildfire.

Prescribed burning for preparing sites for direct seeding was tried on a few occasions in Ontario, but none of the burns was hot enough to produce a seedbed that was adequate without supplementary mechanical site preparation.

Changes in soil chemical properties associated with burning include significantly increased pH, which Macadam (1987) in the Sub-boreal Spruce Zone of central British Columbia found persisting more than a year after the burn. Average fuel consumption was 20 to 24 t/ha and the forest floor depth was reduced by 28% to 36%. The increases correlated well with the amounts of slash (both total and ≥7 cm diameter) consumed. The change in pH depends on the severity of the burn and the amount consumed; the increase can be as much as 2 units, a 100-fold change. Deficiencies of copper and iron in the foliage of white spruce on burned clearcuts in central British Columbia might be attributable to elevated pH levels.

Even a broadcast slash fire in a clearcut does not give a uniform burn over the whole area. Tarrant (1954), for instance, found only 4% of a 140-ha slash burn had burned severely, 47% had burned lightly, and 49% was unburned. Burning after windrowing obviously accentuates the subsequent heterogeneity.

Marked increases in exchangeable calcium also correlated with the amount of slash at least 7 cm in diameter consumed. Phosphorus availability also increased, both in the forest floor and in the 0 cm to 15 cm mineral soil layer, and the increase was still evident, albeit somewhat diminished, 21 months after burning. However, in another study in the same Sub-boreal Spruce Zone found that although it increased immediately after the burn, phosphorus availability had dropped to below pre-burn levels within 9 months.

Nitrogen will be lost from the site by burning, though concentrations in remaining forest floor were found by Macadam (1987) to have increased in 2 of 6 plots, the others showing decreases. Nutrient losses may be outweighed, at least in the short term, by improved soil microclimate through the reduced thickness of forest floor where low soil temperatures are a limiting factor.

The *Picea/Abies* forests of the Alberta foothills are often characterized by deep accumulations of organic matter on the soil surface and cold soil temperatures, both of which make reforestation difficult and result in a general deterioration in site productivity; Endean and Johnstone (1974) describe experiments to test prescribed burning as a means of seedbed preparation and site amelioration on representative clear-felled *Picea/Abies* areas. Results showed that, in general, prescribed burning did not reduce organic layers satisfactorily, nor did it increase soil temperature, on the sites tested. Increases in seedling establishment, survival, and growth on the burned sites were probably the result of slight reductions in the depth of the organic layer, minor increases in soil temperature, and marked improvements in the efficiency of the planting crews. Results also suggested that the process of site deterioration has not been reversed by the burning treatments applied.

Ameliorative Intervention

Slash weight (the oven-dry weight of the entire crown and that portion of the stem < 4 inches in diameter) and size distribution are major factors influencing the forest fire hazard on harvested sites. Forest managers interested in the application of prescribed burning for hazard reduction and silviculture, were shown a method for quantifying the slash load by Kiil (1968). In west-central Alberta, he felled, measured, and weighed 60 white spruce, graphed (a) slash weight per merchantable unit volume against diameter at breast height (dbh), and (b) weight of fine slash (<1.27 cm) also against dbh, and produced a table of slash weight and size distribution on one acre of a hypothetical stand of white spruce. When the diameter distribution of a stand is unknown, an estimate of slash weight and size distribution can be obtained from average stand diameter, number of trees per unit area, and merchantable cubic foot volume. The sample trees in Kiil's study had full symmetrical crowns. Densely growing trees with short and often irregular crowns would probably be overestimated; open-grown trees with long crowns would probably be underestimated.

The need to provide shade for young outplants of Engelmann spruce in the high Rocky Mountains is emphasized by the U.S. Forest Service. Acceptable planting spots are defined as microsites on the north and east sides of down logs, stumps, or slash, and lying in the shadow cast by such material. Where the objectives of management specify more uniform spacing, or higher densities, than obtainable from an existing distribution of shade-providing material, redistribution or importing of such material has been undertaken.

Access

Site preparation on some sites might be done simply to facilitate access by planters, or to improve access and increase the number or distribution of microsites suitable for planting or seeding.

Wang et al. (2000) determined field performance of white and black spruces 8 and 9 years after outplanting on boreal mixedwood sites following site preparation (Donaren disc trenching versus no trenching) in 2 plantation types (open versus sheltered) in southeastern Manitoba. Donaren trenching slightly reduced the mortality of black spruce but significantly increased the mortality of white spruce. Significant difference in height was found between open and sheltered plantations for black spruce but not for white spruce, and root collar diameter in sheltered plantations was significantly larger than in open plantations for black spruce but not for white spruce. Black spruce open plantation had significantly smaller volume (97 cm^3) compared with black spruce sheltered (210 cm^3), as well as white spruce open (175 cm^3) and sheltered (229 cm^3) plantations. White spruce open plantations also had smaller volume than white spruce sheltered plantations. For transplant stock, strip plantations had a significantly higher volume (329 cm^3) than open plantations (204 cm^3). Wang et al. (2000) recommended that sheltered plantation site preparation should be used.

Mechanical

Up to 1970, no "sophisticated" site preparation equipment had become operational in Ontario, but the need for more efficacious and versatile equipment was increasingly recognized. By this time, improvements were being made to equipment originally developed by field staff, and field testing of equipment from other sources was increasing.

According to J. Hall (1970), in Ontario at least, the most widely used site preparation technique was post-harvest mechanical scarification by equipment front-mounted on a bulldozer (blade, rake, V-plow, or teeth), or dragged behind a tractor (Imsett or S.F.I. scarifier, or rolling chopper). Drag type units designed and constructed by Ontario's Department of Lands and Forests used anchor chain or tractor pads separately or in combination, or were finned steel drums or barrels of various sizes and used in sets alone or combined with tractor pad or anchor chain units.

J. Hall's (1970) report on the state of site preparation in Ontario noted that blades and rakes were found to be well suited to post-cut scarification in tolerant hardwood stands for natural regeneration of yellow birch. Plows were most effective for treating dense brush prior to planting, often in conjunction with a planting machine. Scarifying teeth, e.g., Young's teeth, were sometimes used to prepare sites for planting, but their most effective use was found to be preparing sites for seeding, particularly in backlog areas carrying light brush and dense herbaceous growth. Rolling choppers found application in treating heavy brush but could be used only on stone-free soils. Finned drums were commonly used on jack pine–spruce cutovers on fresh brushy sites with a deep duff layer and heavy slash, and they needed to be teamed with a tractor pad unit to secure good distribution of the slash. The S.F.I. scarifier, after strengthening, had been "quite successful" for 2 years, promising trials were under way with the cone scarifier and barrel ring scarifier, and development had begun on a new flail scarifier for use on sites with shallow, rocky soils. Recognition of the need to become more effective and efficient in site preparation led the Ontario Department of Lands and Forests to adopt the policy of seeking and obtaining for field testing new equipment from Scandinavia and elsewhere that seemed to hold promise for Ontario conditions, primarily in the north. Thus, testing was begun of the Brackekultivator from Sweden and the Vako-Visko rotary furrower from Finland.

Mounding

Site preparation treatments that create raised planting spots have commonly improved outplant performance on sites subject to low soil temperature and excess soil moisture. Mounding can certainly have a big influence on soil temperature. Draper et al. (1985), for instance, documented this as well as the effect it had on root growth of outplants (Table 30).

The mounds warmed up quickest, and at soil depths of 0.5 cm and 10 cm averaged 10 and 7 °C higher, respectively, than in the control. On sunny days, daytime surface temperature maxima on the mound and organic mat reached 25 °C to 60 °C, depending on soil wetness and shading. Mounds reached mean soil temperatures of 10 °C at 10 cm depth 5 days after planting, but the control did not reach that temperature until 58 days after planting. During the first growing season, mounds had 3 times as many days with a mean soil temperature greater than 10 °C than did the control microsites.

Draper et al.'s (1985) mounds received 5 times the amount of photosynthetically active radiation (PAR) summed over all sampled microsites throughout the first growing season; the control treatment consistently received about 14% of daily background PAR, while mounds received over 70%. By November, fall frosts had reduced shading, eliminating the differential. Quite apart from its effect on temperature, incident radiation is also important photosynthetically. The average control microsite was exposed to levels of light above the compensation point for only 3 hours, i.e., one-quarter of the daily light period, whereas mounds received light above the compensation point for 11 hours, i.e., 86% of the same daily period. Assuming that incident light in the 100-600 $\mu Em^{-2}s^{-1}$ intensity range is the most important for photosynthesis, the mounds received over 4 times the total daily light energy that reached the control microsites.

Orientation of Linear Site Preparation, e.g., Disk-trenching

With linear site preparation, orientation is sometimes dictated by topography or other considerations, but the orientation can often be chosen. It can make a difference. A disk-trenching experiment in the Sub-boreal Spruce Zone in interior British Columbia investigated the effect on growth of young outplants (lodgepole pine) in 13 microsite planting positions: berm, hinge, and trench in each of north, south, east, and west aspects, as well as in untreated locations between the furrows. Tenth-year stem volumes of trees on south, east, and west-facing microsites were significantly greater than those of trees on north-facing and untreated microsites. However, planting spot selection was seen to be more important overall than trench orientation.

In a Minnesota study, the N–S strips accumulated more snow but snow melted faster than on E–W strips in the first year after felling. Snow-melt was faster on strips near the centre of the strip-felled area than on border strips adjoining the intact stand. The strips, 50 feet (15.24 m) wide, alternating with uncut strips 16 feet (4.88 m) wide, were felled in a *Pinus resinosa* stand, aged 90 to 100 years.

Forest Management

Forest management is a branch of forestry concerned with overall administrative, economic, legal, and social aspects, as well as scientific and technical aspects, such as silviculture, protection, and forest regulation. This includes management for aesthetics, fish, recreation, urban values, water, wilderness, wildlife, wood products, forest genetic resources, and other forest resource values. Management can be based on conservation, economics, or a mixture of the two. Techniques include timber extraction, planting and replanting of various species, cutting roads and pathways through forests, and preventing fire.

Definition

The forest is a natural system that can supply different products and services. The working of this system is influenced by the natural environment: climate, topography, soil, etc., and also by human activity. The actions of humans in forests constitute forest management. In developed societies, this management tends to be elaborated and planned in order to achieve the objectives that are considered desirable.

Some forests have been and are managed to obtain traditional forest products such as firewood, fiber for paper, and timber, with little thinking for other products and services. Nevertheless, as a result of the progression of environmental awareness, management of forests for multiple use is becoming more common.

Public Input and Awareness

There has been increased public awareness of natural resource policy, including forest management. Public concern regarding forest management may have shifted from the extraction of timber for earning money for the economy, to the preservation of additional forest resources, including wildlife and old growth forest, protecting biodiversity, watershed management, and recreation. Increased environmental awareness may contribute to an increased public mistrust of forest management professionals. But it can also lead to greater understanding about what professionals do re forests for nature conservation and ecological services. The importance of taking care of the forests for ecological as well as economical sustainable reasons has been shown in the TV show Ax Men.

FSC

The ecolabel of the Forest Stewardship Council for certified sustainable forest product

Many tools like GIS and photogrammetry modelling have been developed to improve forest inventory and management planning.

Wildlife Considerations

The abundance and diversity of birds, mammals, amphibians and other wildlife are affected by strategies and types of forest management.

Management Intensity

Forest management varies in intensity from a leave alone, natural situation to a highly intensive regime with silvicultural interventions. Management is generally increased in intensity to achieve either economic criteria (increased timber yields, non-timber forest products, ecosystem services) or ecological criteria (species recovery, fostering of rare species, carbon sequestration).

Forest Pathology

Forest pathology is the research of both biotic and abiotic maladies affecting the health of a forest ecosystem, primarily fungal pathogens and their insect vectors. It is a subfield of forestry and plant pathology.

Forest pathology is part of the broader approach of forest protection.

Abiotic factors

There are a number of abiotic factors which affect the health of a forest, such as moisture issues like drought, winter-drying, waterlogging resulting from over-abundance or lack of precipitation such as hail, snow, rain.

Wind is also an important abiotic factor as windthrow (the uprooting or breaking of trees due to high winds) causes an obvious and direct loss of stability to a forest or its trees.

Often, abiotic factors and biotic factors will affect a forest at the same time. For example, if wind speed is 80 km per hour then many trees which have root rot (caused by a pathogen) are likely to be thrown. Higher wind speeds are necessary to damage healthier trees.

Fire, whether caused by humans or lightning and related abiotic factors also affect the health of forest.

The effects of man often alter a forest's predisposition to damage from both abiotic and biotic effects. For example soil properties may be altered by heavy machinery.

Other abiotic factors

- Nutrient imbalances: deficiencies, chemicals (toxic salts, herbicides, air pollutants)
- Stemflow which can concentrate dry deposits which via soil acidification can kill surrounding plants.
- Temperature

Some of these factors act in concert (all do to a degree). For example Amylostereum areolatum which is spread by the Sirex woodwasp. The fungus gains access to new trees to live off and the woodwasp larvae gain food.

Parasitic Flowering Plants

- Many plants can parasitize trees via root to root contact. Many of these parasitic plants originate in the tropical and subtropical climates.

Animals

Nematodes, insects especially bark beetles, mammals may browse. Browsing can be prevented with tree shelters

Part of forest pathology is forest entomology. Forest entomology includes the study of all insects and arthropods, such as mites, centipedes and millipedes, which live in and interact in forest ecosystems. Forest entomology also includes the management of insect pests that cause the degrading, defoliation, crown die-back or death of trees.

Thus the scope is wide and includes:

- Documentation of all insect species and related arthropoda in natural and man-made forests, and the study and ecology of those species.
- Description and assessment of damage to tree structures (parts of a tree), to forest stands, landscape effects and to wood products, timber in service and other ecosystem services.
- Eradication of recently introduced pests, or long-term management of established exotics and indigenous pests, to minimise losses in wood quality and wood production, and to reduce tree mortality.
- Assessments of forest operations, or of management impacts, on the invertebrate fauna, and the alleviation of any adverse effects on these invertebrates.

Hazard Trees

- The likelihood of property damage or personal injury due to tree failure. Hazard includes not only the tree's condition, but the potential target as well. Rating systems, procedures and guidelines have been developed for decision making but knowledge, judgement, and experience are an important part to the process.

Signs and Symptoms

Symptoms are a result of a pathogen:

- Blight
- Burl
- Canker
- Chlorosis
- Drunken trees
- Forest dieback
- Gall
- Girdling
- Leaf scorch
- Root rot
- Virescence

- Wilt disease

Signs are the visible presence of a part of a pathogen:

- Ascus is a part of an acomycota fungus.
- Conk (fungi) is the fruiting body of a bracket fungus.
- Hypha are collectively called a mycelium
- Mycelial cord or rhizomorphs

Pathology Detection

This can be done by dogs or machines smelling the trees. It is similar to noses used to find truffles. It can also be done by monitoring and identification can happen via tree clinics, experts such as arborists or even non-experts through citizen science.

It is important to consider the disease triangle when evaluating pathologies. Demonstration of suspected active agents can be done by confirmation of Koch's postulates.

Hardwood Timber Production

Hardwood timber production is the process of managing stands of deciduous trees to maximize woody output. The production process is not linear because other factors must be considered, including marketable and non-marketable goods, financial benefits, management practices, and the environmental implications of those management practices.

Forests include market and non-market products. Marketable products include goods that have a market price. Timber is the main one, with prices that range from a few hundred dollars per thousand board feet (MBF) to several thousand dollars for a veneer log. Others include grazing/fodder, specialty crops such as mushrooms or berries, usage fees for recreation or hunting, and biomass. Forests also provide some non-market values which have no current market price. Examples of non-market goods would be improving water quality, air quality, aesthetics, and carbon sequestration.

The more biodiverse the hardwood-forest ecosystem, the more challenges and opportunities its managers face. Managers aim for sustainable forest management to keep their cash crop renewing itself, using silvicultural practices that include harvesting, promoting regeneration, controlling insects and disease, fertilizing, applying herbicide treatments, and thinning. Fertilization can increase the growth rate and amount of plant material, thus possibly increasing the number of wildlife that can inhabit a site. Invasive species control maintains an area's structure and native composition.

But management can also harm the ecosystem; for example, machinery used in a timber harvest can compact the soil, stress the root system, reduce tree growth, lengthen the time needed for a stand to mature to harvestability. Machinery can also damage the understory, disturbing wildlife habitat and prevent regeneration.

Arboriculture

An arborist practicing arboriculture: using a chainsaw to fell a eucalyptus tree in a park at Kallista, Victoria.

Arboriculture is the cultivation, management, and study of individual trees, shrubs, vines, and other perennial woody plants. The science of arboriculture studies how these plants grow and respond to cultural practices and to their environment. The practice of arboriculture includes cultural techniques such as selection, planting, training, fertilization, pest and pathogen control, pruning, shaping, and removal.

A person who practices or studies arboriculture can be termed an 'arborist' or an 'arboriculturist'. A 'tree surgeon' is more typically someone who is trained in the physical maintenance and manipulation of trees and therefore more a part of the arboriculture process rather than an arborist. Risk management, legal issues, and aesthetic considerations have come to play prominent roles in the practice of arboriculture. Businesses often need to hire arboriculturists to complete "tree hazard surveys" and generally manage the trees on-site to fulfill occupational safety and health obligations.

Arboriculture is primarily focused on individual woody plants and trees maintained for permanent landscape and amenity purposes, usually in gardens, parks or other populated settings, by arborists, for the enjoyment, protection, and benefit of human beings. It falls under the general umbrella of horticulture.

Interception (Water)

Some rain collecting on an acer leaf.

Interception refers to precipitation that does not reach the soil, but is instead intercepted by the leaves and branches of plants and the forest floor. It occurs in the canopy (i.e. canopy interception), and in the forest floor or litter layer (i.e. forest floor interception). Because of evaporation, interception of liquid water generally leads to loss of that precipitation for the drainage basin, except for cases such as fog interception.

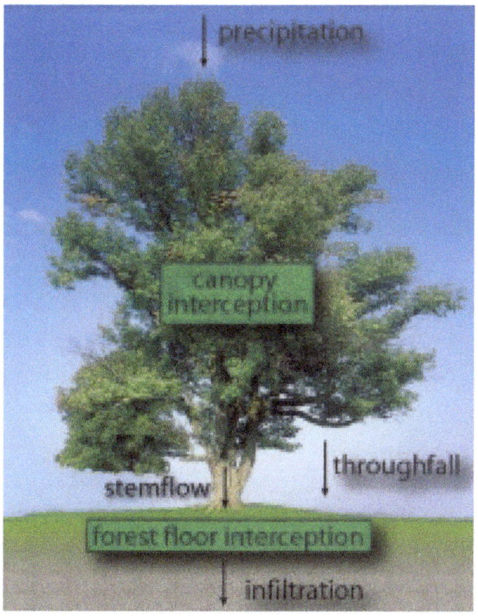

Definition of canopy and forest floor interception

Intercepted snowfall does not result in any notable amount of evaporation, and most of the snow falls off the tree by wind or melts. However, intercepted snow can more easily drift with the wind, out of the watershed. Conifers have a greater interception capacity than hardwoods. Their needles gives them more surface area for droplets to adhere to, and they have foliage in spring and fall, therefore interception also depends on the type of vegetation in a wooded area.

Mitscherlich in 1971 calculated the water storage potential as interception values for different species and stand densities. A storm event might produce 50 - 100 mm of rainfall and 4 mm might be the maximum intercepted in this way. Grah and Wilson in 1944 did sprinkling experiments where they watered plants to see how much of the intercepted is kept after watering stops.

The interception depends on the leaf area index and what kind of leaves they are.

Interception may increase erosion or reduce it depending on the throughfall effects.

Forest Transition

Forest transition refers to a geographic theory describing a reversal or turnaround in land-use trends for a given territory from a period of net forest area loss (i.e., deforestation) to a period of net forest area gain. The term "landscape turnaround" has also been used to represent a more general recovery of natural areas that is independent of biome type.

Forest recovery resulting in net increases in forest extent can occur by means of spontaneous regeneration, active planting, or both.

Forest transitions are associated with socio-economic transformations towards increased industrialization and urbanization. Other conditions leading to the abandonment of agricultural land (e.g., war and environmental legislation) have been found to play important roles in some cases. The different processes through which forest transitions occur are contingent upon the local socioeconomic and ecological contexts. Although some generic processes can be identified, countries do not necessarily experience a regular pattern of forest cover changes with time or development, and the causes and outcomes of forest transitions vary.

Studies of forest transitions have been conducted for several nations as well as sub-national regions. Territories reported to have experienced forest transitions after the onset of industrialization include Bangladesh, China, Costa Rica, Cuba, Denmark,Canada, Dominican Republic, El Salvador, France,Greece, Gambia, Hungary, Ireland, Italy, Japan, Morocco, Norway, New Zealand, Portugal, Puerto Rico, Rwanda, Scotland, South Korea, Spain, Switzerland, the United States, the United Kingdom, and Vietnam. Furthermore, forest-transition dynamics have been documented for regions within Brazil, Ecuador, and Mexico.

The environmental effects of these forest transitions are very variable, depending on whether deforestation of old-growth forests continue, the proportions and types of tree plantations versus natural regeneration of forests, and the location and spatial configuration of the different types of forests.

The findings of returning forests in these widespread studies raise questions about the prospects of a worldwide forest transition, particularly given ongoing processes of forest loss in many parts of the world. In other words, can the global extent of forests be expected to reach a turning point in the future, reversing the current trend of overall forest decline towards overall forest expansion? Studies showed that given an increased competition for productive land between different land uses, a global forest transition would require major policy and technological innovations as well as shifts in demands for fiber, fuel, and food, and that these changes cannot be taken for granted.

Sustainable Forest Management

Sustainable forest management (SFM) is the management of forests according to the principles of sustainable development. Sustainable forest management has to keep the balance between three main pillars: ecological, economic and socio-cultural. Successfully achieving sustainable forest management will provide integrated benefits to all, ranging from safeguarding local livelihoods to protecting the biodiversity and ecosystems provided by forests, reducing rural poverty and mitigating some of the effects of climate change.

The "Forest Principles" adopted at The United Nations Conference on Environment and Development (UNCED) in Rio de Janeiro in 1992 captured the general international understanding of sustainable forest management at that time. A number of sets of criteria and indicators have since been developed to evaluate the achievement of SFM at the global, regional, country and management unit level. These were all attempts to codify and provide for independent assessment of the degree to which the broader objectives of sustainable forest management are being achieved in practice. In 2007, the United Nations General Assembly adopted the Non-Legally Binding Instrument on All Types of Forests. The instrument was the first of its kind, and reflected the strong international commitment to promote implementation of sustainable forest management through a new approach that brings all stakeholders together.

Definition

A definition of SFM was developed by the Ministerial Conference on the Protection of Forests in Europe (FOREST EUROPE), and has since been adopted by the Food and Agriculture Organization (FAO). It defines sustainable forest management as:

The stewardship and use of forests and forest lands in a way, and at a rate, that maintains their biodiversity, productivity, regeneration capacity, vitality and their potential to fulfill, now and in the future, relevant ecological, economic and social functions, at local, national, and global levels, and that does not cause damage to other ecosystems.

In simpler terms, the concept can be described as the attainment of balance – balance between society's increasing demands for forest products and benefits, and the preservation of forest health and diversity. This balance is critical to the survival of forests, and to the prosperity of forest-dependent communities.

For forest managers, sustainably managing a particular forest tract means determining, in a tangible way, how to use it today to ensure similar benefits, health and productivity in the future. Forest managers must assess and integrate a wide array of sometimes conflicting factors – commercial and non-commercial values, environmental considerations, community needs, even global impact – to produce sound forest plans. In most cases, forest managers develop their forest plans in consultation with citizens, businesses, organizations and other interested parties in and around the forest tract being managed. The tools and visualization have been recently evolving for better management practices.

In 2014, at the request of Member States, the Food and Agriculture Organization of the United Nations developed and launched the Sustainable Forest Management Toolbox, an online collection of tools, case studies and other resources to support countries implementing sustainable forest management.

Because forests and societies are in constant flux, the desired outcome of sustainable forest management is not a fixed one. What constitutes a sustainably managed forest will change over time as values held by the public change.

Criteria and Indicators

Deforestation of native rain forest in Rio de Janeiro City for extraction of clay for civil engineering (2009 picture). An example of non sustainable forest management.

Criteria and indicators are tools which can be used to conceptualise, evaluate and implement sustainable forest management. Criteria define and characterize the essential elements, as well as a set of conditions or processes, by which sustainable forest management may be assessed. Periodically measured indicators reveal the direction of change with respect to each criterion.

Criteria and indicators of sustainable forest management are widely used and many countries produce national reports that assess their progress toward sustainable forest management. There are nine international and regional criteria and indicators initiatives, which collectively involve more than 150 countries. Three of the more advanced initiatives are those of the Working Group on Criteria and Indicators for the Conservation and Sustainable Management of Temperate and Boreal Forests (also called the Montreal Process), Forest Europe, and the International Tropical Timber Organization. Countries who are members of the same initiative usually agree to produce reports at the same time and using the same indicators. Within countries, at the management unit level, efforts have also been directed at developing local level criteria and indicators of sustainable forest management. The Center for International Forestry Research, the International Model Forest Network and researchers at the University of British Columbia have developed a number of tools and techniques to help forest-dependent communities develop their own local level criteria and indicators. Criteria and Indicators also form the basis of third-party forest certification programs such as the Canadian Standards Association's Sustainable Forest Management Standards and the Sustainable Forestry Initiative Standard.

There appears to be growing international consensus on the key elements of sustainable forest management. Seven common thematic areas of sustainable forest management have emerged based on the criteria of the nine ongoing regional and international criteria and indicators initiatives. The seven thematic areas are:

- Extent of forest resources
- Biological diversity
- Forest health and vitality
- Productive functions and forest resources
- Protective functions of forest resources
- Socio-economic functions
- Legal, policy and institutional framework.

This consensus on common thematic areas (or criteria) effectively provides a common, implicit definition of sustainable forest management. The seven thematic areas were acknowledged by the international forest community at the fourth session of the United Nations Forum on Forests and the 16th session of the Committee on Forestry. These thematic areas have since been enshrined in the Non-Legally Binding Instrument on All Types of Forests as a reference framework for sustainable forest management to help achieve the purpose of the instrument.

On January 5, 2012, the Montreal Process, Forest Europe, the International Tropical Timber Organization, and the Food and Agriculture Organization of the United Nations, acknowledging the seven thematic areas, endorsed a joint statement of collaboration to improve global forest related data collection and reporting and avoiding the proliferation of monitoring requirements and associated reporting burdens.

Ecosystem Approach

The Ecosystem Approach has been prominent on the agenda of the Convention on Biological Diversity (CBD) since 1995 . The CBD definition of the Ecosystem Approach and a set of principles for its application were developed at an expert meeting in Malawi in 1995, known as the Malawi Principles. The definition, 12 principles and 5 points of "operational guidance" were adopted by the fifth Conference of Parties (COP5) in 2000. The CBD definition is as follows

The ecosystem approach is a strategy for the integrated management of land, water and living resources that promotes conservation and sustainable use in an equitable way. Application of the ecosystem approach will help to reach a balance of the three objectives of the Convention. An ecosystem approach is based on the application of appropriate scientific methodologies focused on levels of biological organization, which encompasses the essential structures, processes, functions and interactions among organisms and their environment. It recognizes that humans, with their cultural diversity, are an integral component of many ecosystems.

Sustainable forest management was recognized by parties to the Convention on Biological Diversity in 2004 (Decision VII/11 of COP7) to be a concrete means of applying the Ecosystem Approach to forest ecosystems. The two concepts, sustainable forest management and the ecosystem approach, aim at promoting conservation and management practices which are environmentally, socially and economically sustainable, and which generate and maintain benefits for both present and future generations. In Europe, the MCPFE and the Council for the Pan-European Biological and Landscape Diversity Strategy (PEBLDS) jointly recognized sustainable forest management to be consistent with the Ecosystem Approach in 2006.

Independent Certification

Growing environmental awareness and consumer demand for more socially responsible businesses helped third-party forest certification emerge in the 1990s as a credible tool for communicating the environmental and social performance of forest operations.

There are many potential users of certification, including: forest managers, scientists, policy makers, investors, environmental advocates, business consumers of wood and paper, and individuals.

With third-party forest certification, an independent organization develops standards of good forest management, and independent auditors issue certificates to forest operations that comply with those standards. Forest certification verifies that forests are well-managed—as defined by a particular standard—and chain-of-custody certification tracks wood and paper products from the certified forest through processing to the point of sale.

This rise of certification led to the emergence of several different systems throughout the world. As a result, there is no single accepted forest management standard worldwide, and each system takes a somewhat different approach in defining standards for sustainable forest management.

In its 2009–2010 Forest Products Annual Market Review United Nations Economic Commission for Europe/Food and Agriculture Organization stated: "Over the years, many of the issues that previously divided the (certification) systems have become much less distinct. The largest certification systems now generally have the same structural programmatic requirements. "

Third-party forest certification is an important tool for those seeking to ensure that the paper and wood products they purchase and use come from forests that are well-managed and legally harvested. Incorporating third-party certification into forest product procurement practices can be a centerpiece for comprehensive wood and paper policies that include factors such as the protection of sensitive forest values, thoughtful material selection and efficient use of products.

There are more than 50 certification standards worldwide, addressing the diversity of forest types and tenures. Globally, the two largest umbrella certification programs are:

- Programme for the Endorsement of Forest Certification (PEFC)
- Forest Stewardship Council (FSC)

The area of forest certified worldwide is growing slowly. PEFC is the world's largest forest certification system, with more than two-thirds of the total global certified area certified to its Sustainability Benchmarks.

In North America, there are three certification standards endorsed by PEFC – the Sustainable Forestry Initiative, the Canadian Standards Association's Sustainable Forest Management Standard, and the American Tree Farm System. FSC has five standards in North America – one in the United States and four in Canada.

While certification is intended as a tool to enhance forest management practices throughout the world, to date most certified forestry operations are located in Europe and North America. A significant barrier for many forest managers in developing countries is that they lack the capacity to undergo a certification audit and maintain operations to a certification standard.

Forest Governance

Countries participating in the UNREDD program and/or Forest Carbon Partnership Facility.

UN-REDD participants

Forest Carbon Partnership Facility participants

participants in both

Although a majority of forests continue to be owned formally by government, the effectiveness of forest governance is increasingly independent of formal ownership. Since neo-liberal ideology in the 1980s and the emanation of the climate change challenges, evidence that the state is failing to effectively manage environmental resources has emerged. Under neo-liberal regimes in the developing countries, the role of the state has diminished and the market forces have increasingly taken over the dominant socio-economic role. Though the critiques of neo-liberal policies have maintained that market forces are not only inappropriate for sustaining the environment, but are in fact a major cause of environmental destruction. Hardin's tragedy of the common (1968) has shown that the people cannot be left to do as they wish with land or environmental resources. Thus, decentralization of management offers an alternative solution to forest governance.

The shifting of natural resource management responsibilities from central to state and local governments, where this is occurring, is usually a part of broader decentralization process. According to Rondinelli and Cheema (1983), there are four distinct decentralization options: these are: (i) Privatization – the transfer of authority from the central government to non-governmental sectors otherwise known as market-based service provision, (ii) Delegation – centrally nominated local authority, (iii) Devolution – transfer of power to locally acceptable authority and (iv) Deconcentration – the redistribution of authority from the central government to field delegations of the central government. The major key to effective decentralization is increased broad-based participation in local-public decision making. In 2000, the World Bank report reveals that local government knows the needs and desires of their constituents better than the national government, while at the same time, it is easier to hold local leaders accountable. From the study of West African tropical forest, it is argued that the downwardly accountable and/or representative authorities with meaningful discretional powers are the basic institutional element of decentralization that should lead to efficiency, development and equity. This collaborates with the World Bank report in 2000 which says that decentralization should improve resource allocation, efficiency, accountability and equity "by linking the cost and benefit of local services more closely".

Many reasons point to the advocacy of decentralization of forest. (i) Integrated rural development projects often fail because they are top-down projects that did not take local people's needs and desires into account. (ii) National government sometimes have legal authority over vast forest areas that they cannot control, thus, many protected area projects result in increased biodiversity loss and greater social conflict. Within the sphere of forest management, as state earlier, the most effective option of decentralization is "devolution"-the transfer of power to locally accountable authority. However, apprehension about local governments is not unfounded. They are often short of resources, may be staffed by people with low education and are sometimes captured by local elites who promote clientelist relation rather than democratic participation. Enters and Anderson (1999) point that the result of community-based projects intended to reverse the problems of past central approaches to conservation and development have also been discouraging.

Broadly speaking, the goal of forest conservation has historically not been met when, in contrast with land use changes; driven by demand for food, fuel and profit. It is necessary to recognize and advocate for better forest governance more strongly given the importance of forest in meeting basic human needs in the future and maintaining ecosystem and biodiversity as well as addressing climate change mitigation and adaptation goal. Such advocacy must be coupled with financial incentives for government of developing countries and greater governance role for local government, civil society, private sector and NGOs on behalf of the "communities".

National Forest Funds

The development of National Forest Funds is one way to address the issue of financing sustainable forest management. National forest funds (NFFs) are dedicated financing mechanisms managed by public institutions designed to support the conservation and sustainable use of forest resources. As of 2014, there are 70 NFFs operating globally.

Dendrology

Dendrology (*dendron*, "tree"; and *-logia*, *science of* or *study of*) or xylology (*ksulon*, "wood") is the science and study of wooded plants (trees, shrubs, and lianas), specifically, their taxonomic classifications. There is no sharp boundary between plant taxonomy and dendrology; however, woody plants not only belong to many different plant families, but these families may be made up of both woody and non-woody members. Some families include only a few woody species. Dendrology, as a discipline of industrial forestry, tends to focus on identification of economically useful woody plants and their taxonomic interrelationships. As an academic course of study, Dendrology will include all woody plants, native and non-native, that occur in a region. A related discipline is the study of Sylvics, which focuses on the autecology of genera and species.

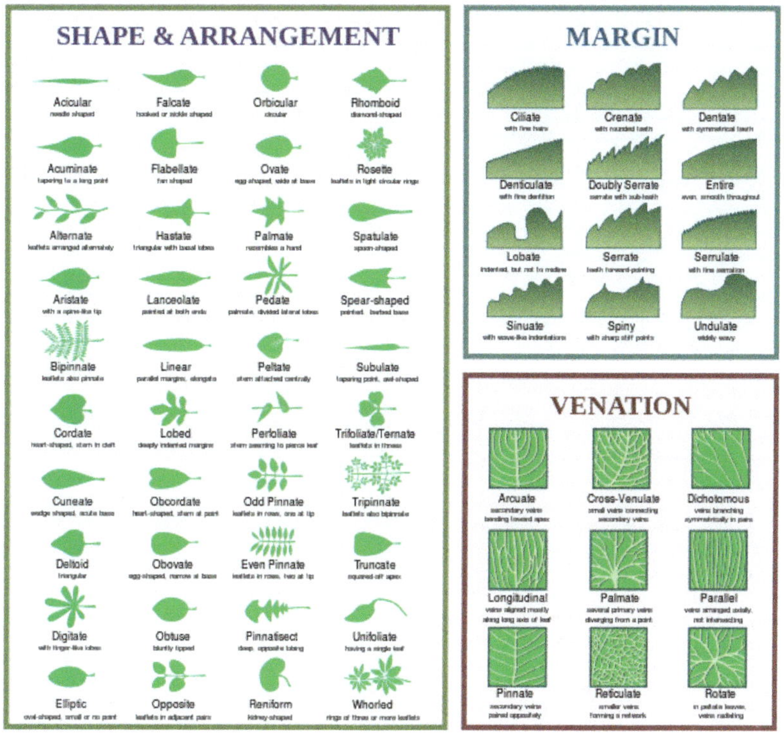

Leaf shape is a common method used to identify trees.

Relationship with Botany

Dendrology is often confused with botany. However, botany is the study of all types of general plants, while dendrology studies only woody plants. Dendrology may be considered a subcategory of botany that specializes in the characterization and identification of woody plants.

References

- Kohm, K. A, and Franklin, J. F., Creating a forestry for the 21st century: the science of ecosystem management. Island Press. 1997, ISBN 978-1-55963-399-4
- Harris, Richard W. (1983). Arboriculture: Care of Trees, Shrubs, and Vines in the Landscape. Englewood Cliffs, New Jersey 07632: Prentice-Hall, Inc. pp. 2–3. ISBN 0-13-043935-5.
- Matta, Rao (2015). Towards effective national forest funds, FAO Forestry Paper 174 (PDF). Rome, Italy: Food and Agriculture Organization of the United Nations. ISBN 978-92-5-108706-0.
- Matta, Rao (2015). Towards effective national forest funds, FAO Forestry Paper 174 (PDF). Rome, Italy: Food and Agriculture Organization of the United Nations. ISBN 978-92-5-108706-0.

- "LEDS GP Agriculture, Forestry and Other Land Use Working Group factsheet" (PDF). Low Emission Development Strategies Global Partnership (LEDS GP). Retrieved 23 March 2016.
- "Evans, K., De Jong, W., and Cronkleton, P. (2008) "Future Scenarios as a Tool for Collaboration in Forest Communities". ''S.A.P.I.EN.S.'' '''1''' (2)". Sapiens.revues.org. 1 October 2008. Retrieved 30 November 2011.
- "Mozgeris, G. (2008) "The continuous field view of representing forest geographically: from cartographic representation towards improved management planning". ''S.A.P.I.EN.S.'' '''1''' (2)". Sapiens.revues.org. Retrieved 30 November 2011.

3

Forestry Practices

One of the most important practices involved in forestry is logging and tree cutting for timber. Subjects involved in this chapter include various types of logging, rubber tapping, clear cutting and plant nurseries. These practices maintain the balance between timber being felled and new saplings that are being cultivated. The chapter strategically encompasses and incorporates the major components and key concepts of forestry, providing a complete understanding.

Logging

A mountain ash (*Eucalyptus regnans*) being felled using springboards, c. 1884–1917, Australia

Logging is the cutting, skidding, on-site processing, and loading of trees or logs onto trucks or skeleton cars.

In forestry, the term logging is sometimes used in a narrow sense concerning the logistics of moving wood from the stump to somewhere outside the forest, usually a sawmill or a lumber yard. However, in common usage, the term may be used to indicate a range of forestry or silviculture activities.

Illegal logging refers to what in forestry might be called timber theft by the timber mafia. It can also refer to the harvesting, transportation, purchase, or sale of timber in violation of laws. The harvesting procedure itself may be illegal, including using corrupt means to gain access to forests;

extraction without permission or from a protected area; the cutting of protected species; or the extraction of timber in excess of agreed limits.

McGiffert Log Loader in East Texas, USA circa 1907

Clearcut logging is not necessarily considered a type of logging but a harvesting or silviculture method, and is simply called clearcutting or block cutting. In the forest products industry logging companies may be referred to as logging contractors, with the smaller, non-union crews referred to as "gyppo loggers."

Cutting trees with the highest value and leaving those with lower value, often diseased or malformed trees, is referred to as high grading. It is sometimes called selective logging, and confused with selection cutting, the practice of managing stands by harvesting a proportion of trees.

Logging usually refers to above-ground forestry logging. Submerged forests exist on land that has been flooded by damming to create reservoirs. Such trees are logged using underwater logging or by the lowering of the reservoirs in question. Ootsa Lake and Williston Lake in British Columbia, Canada are notable examples where timber recovery has been needed to remove inundated forests.

Clearcutting

Log transportation

Clearcutting, or clearfelling, is a method of harvesting that removes essentially all the standing trees in a selected area. Depending on management objectives, a clearcut may or may not have reserve trees left to attain goals other than regeneration, including wildlife habitat management, mitigation of potential erosion or water quality concerns. Silviculture objectives for clearcutting, (for example, healthy regeneration of new trees on the site) and a focus on forestry distinguish it from deforestation. Other methods include shelterwood cutting, group selective, single selective, seed-tree cutting, patch cut, and retention cutting.

Logging Methods

The above operations can be carried out by different methods, of which the following three are considered industrial methods:

The Washington Iron Works Skidder in Nuniong is the only one of its kind in Australia, with donkey engine, spars, and cables still rigged for work.

Tree-length Logging / Stem Only Harvesting (SOH)

Trees are felled and then delimbed and topped at the stump. The log is then transported to the landing, where it is bucked and loaded on a truck. This leaves the slash (and the nutrients it contains) in the cut area, where it must be further treated if wild land fires are of concern.

Whole-Tree Logging (WTL)

Horse logging in Poland

Cable logging in French Alps (cable grue Larix 3T)

Logging in Hungary

Trees and plants are felled and transported to the roadside with top and limbs intact.There have been advancements to the process which now allows a logger or harvester to cut the tree down, top, and delimb a tree in the same process. This ability is due to the advancement in the style felling head that can be used. The trees are then delimbed, topped, and bucked at the landing. This method requires that slash be treated at the landing. In areas with access to cogeneration facilities, the slash can be chipped and used for the production of electricity or heat. Full-tree harvesting also refers to utilization of the entire tree including branches and tops. This technique removes both nutrients and soil cover from the site and so can be harmful to the long term health of the area if no further action is taken, however, depending on the species, many of the limbs are often broken off in handling so the end result may not be as different from tree-length logging as it might seem.

Cut-to-length Logging

Cut-to-length logging is the process of felling, delimbing, bucking, and sorting (pulpwood, sawlog, etc.) at the stump area, leaving limbs and tops in the forest. Harvesters fell the tree, delimb, and buck it, and place the resulting logs in bunks to be brought to the landing by a skidder or forwarder. This method is routinely available for trees up to 900 mm (35 in) in diameter. Harvesters are employed effectively in level to moderately steep terrain. Harvesters are highly computerized to optimize cutting length, control harvesting area by GPS, and utilize price lists for each specific log to archive most economical results during harvesting.

Hardwood logs transported down the Suriname River, Suriname, South America in 1955

Transporting Logs

Timber rafting in Joensuu canal in 2009

Log transport in the Dutch East Indes (now Indonesia) c. 1870.

Paperwood awaiting transfer

Timberjack Harvester at work

Felled logs are then generally transported to a sawmill to be cut into lumber, to a paper mill for paper pulp, or for other uses, for example, as fence posts. Many methods have been used to move logs from where they were cut to a rail line or directly to a sawmill or paper mill. The cheapest and historically most common method is making use of a river's current to float floating tree trunks downstream, by either log driving or timber rafting. (Some logs sink because of high resin content; these are called deadheads.) To help herd the logs to the mill, in 1960 the Alaskan Lumber and Pulp Mill had a specially designed boat that was constructed of 1 $\frac{1}{2}$ inch (38 mm) steel. In the late 1800s and the first half of the 1900s, the most common method was the high-wheel loader, which was a set of wheels over ten feet tall that the log or logs were strapped beneath. Oxen were at first used with the high-wheel loaders, but in the 1930s tractors replaced the oxen. In 1960 the largest high wheel loader was built for service in California. Called the Bunyan Buggie, the unit was self-propelled and had wheels 24 feet (7.3 m) high and a front dozer blade that was 30 feet (9.1 m) across and 6 feet (1.8 m) high. Log transportation can be challenging and costly since trees are often far from roads or watercourses. Road building and maintenance may be restricted in National Forests or other wilderness areas since it can cause erosion in riparian zones. When felled logs sit adjacent to a road, heavy machinery may simply lift logs onto trucks. Most often, special heavy equipment is used to gather the logs from the site and move them close to the road to be lifted on trucks. Many methods exist to transport felled logs lying away from roads. Cable logging involves a yarder, which pulls one or several logs along the ground to a platform where a truck is waiting. When the terrain is too uneven to pull logs on the ground, a skyline can lift logs off the ground vertically, similar to a ski lift. Heli-logging, which uses heavy-lift helicopters to remove cut trees from forests by lifting them on cables attached to a helicopter, may be used when cable logging is not allowed for environmental reasons or when roads are lacking. It reduces the level of infrastructure required to log in a specific location, reducing the environmental impact of logging. Less mainstream or now for the most part superseded forms of log transport include horses, oxen, or balloon logging.

Safety Considerations

Logging is a dangerous occupation. In the United States, it has consistently been one of the most hazardous industries. In 2008, the logging industry employed 86,000 workers, and accounted for 93 deaths. This resulted in a fatality rate of 108.1 deaths per 100,000 workers that year. This rate is over 30 times higher than the overall fatality rate. Loggers work with heavy, moving weights, and use tools such as chainsaws and heavy equipment on uneven and sometimes steep or unstable terrain. Loggers also deal with severe environmental conditions, such as inclement weather and severe heat or cold. An injured logger is often far from professional emergency treatment.

Traditionally, the cry of "Timber!" developed as a warning alerting fellow workers in an area that a tree is being felled, so they should be alert to avoid being struck. The term "widowmaker" for timber that is neither standing nor fallen to the ground demonstrates another emphasis on situational awareness as a safety principle.

In British Columbia, Canada, the BC Forest Safety Council was created in September 2004 as a not-for-profit society dedicated to promoting safety in the forest sector. It works with employers, workers, contractors, and government agencies to implement fundamental changes necessary to make it safer to earn a living in forestry.

The risks experienced in logging operations can be somewhat reduced, where conditions permit, by the use of mechanical tree harvesters, skidders, and forwarders.

Madagascar Rosewood Logging Controversy

In 2009 controversy arose surrounding the intensification of rosewood logging in Madagascar's national parks. Logging was linked to criminal syndicates that laundered rosewood logs through Reunion and Mauritius before transporting timber to China for processing. Finished wood and furniture was then shipped to Europe and the United States. In November 2009 Gibson Guitar Corporation in Nashville was raided by federal authorities for its alleged use of illegally sourced rosewood. The case was settled on August 6, 2012, with Gibson admitting to violating the Lacey Act and agreeing to pay a fine of $300,000 in addition to a $50,000 community payment. Gibson also forfeited the wood seized in the raids, which was valued at roughly the same amount as the settlement. In March 2010 the Malagasy government finally announced a ban of the rosewood export for a period between 2 and 5 years, while as of December 2015 Ralava Beboarimisa the Minister of Environment, Ecology, Sea and Forests is working on a case where rosewood were in transit at Singapore and bound for Hong Kong.

Logging Practies

Log Bucking

Bucker measuring and swamping or knot bumping

Bucking is the process of cutting a felled and delimbed tree into logs. This can be a complicated process because logs destined for plywood, lumber, and pulp each have their own price and specifications for length, diameter, and defects. Significant value can be lost by sub-optimal bucking. Cutting from the top down is *overbucking* and from the bottom up is *underbucking*.

Bucker - Making the Cut

Methods

A felled and delimbed tree has to be cut into logs of standard sizes. This process is called *bucking*, and logger who specialises in this job is a *buck sawyer*.

Bucking may be done in a variety of ways depending on the logging operation. Trees that have been previously felled and moved to a landing with a log skidder are spread out for processing.

While many of the limbs have broken off during transport, the remaining limbs and stubs have to be trimmed. The bucker will anchor the end of an auto rewinding tape measure which is attached to his belt and walk down the log trimming as he goes. The tape is anchored gently with a bent horseshoe nail in the bark so it can be jerked loose when the measurement is completed. When a suitable place to buck the tree is located the cut is made. Local market conditions will determine the particular lengths cut. It is common for log buyers to issue purchase orders for the length, diameter, grade, and species that they prefer and will accept. On the West Coast common cuts on a large Pine or Fir tree is three 32's and a 10. There are often different prices for different items. The bucker is the one who turns a tree into logs, and to do his job properly must cut the tree for optimum value. Significant value may be lost by sub-optimal bucking.

The person bucking is generally called a *bucksawyer* or *bucklogger*, or just a *bucker* and runs as many saws as he can, switching out saws as soon as one is dull. The reason for this is the bucksawyer is typically paid per section of log he cuts. Generally you will find a bucklogger at the smaller sawmills that aren't fully mechanized. This method of logging is perhaps more dangerous than the actual felling of the trees for the bucklogger is usually cutting from the edge of a treepile which can be twenty feet high and as long as there is room to dump them from the truck. Each tree has to be picked out of the pile and cut so that a controlled fall of more trees can be worked as the former fall has been cut and skidded to its respective pile.

Bolts, Billets and Rounds

The pieces of bucked logs may be known by several names. *Bolts* are the pieces of a log which has been bucked into specific lengths which are less than 8 feet (2.4 m), especially short lengths. The etymology of *bolt* is related to being short and stout and related to *knock*, and *strike* possibly because bolts were traditionally split into wood shingles, treenails, clapboards, etc. These pieces may be more specifically known as peeler, shingle, stave or pulpwood bolts. *Billet* is variously defined as a short piece of *round* or partially round wood usually a smaller diameter than a *block* or bolt or as a piece split or cut from a bolt, or sometimes synonymous with *bolt*, particularly when the pieces are intended as firewood (fuel) and sometimes means a piece of a billet after it has been split. *Round* is often associated with lengths of un-split firewood.

Log Driving

Log drivers at Klarälven in Sweden.

Log driving is a means of moving logs (sawn tree trunks) from a forest to sawmills and pulp mills downstream using the current of a river. It was the main transportation method of the early logging industry in Europe and North America.

History

When the first sawmills were established, they were usually small water powered facilities located near the source of timber, which might be converted to grist mills after farming became established when the forests had been cleared. Later, larger circular sawmills were developed in the lower reaches of a river, with the logs floated down to them by log drivers. In the broader, slower stretches of a river, the logs might be bound together into timber rafts. In the smaller, wilder stretches of a river, rafts couldn't get through, so masses of individual logs were driven down the river like huge herds of cattle.

Log jam at Berlin, New Hampshire.

The log drive was one step in a larger process of lumber-making in remote places. In a location with snowy winters, the yearly process typically began in the fall when a small team of men hauled tools upstream into the timbered area, chopped out a clearing, and constructed crude buildings for a logging camp. In the winter when things froze down, a larger crew moved into the camp and proceeded to fell trees, cutting the trunks into 16-foot (5 m) lengths, and hauling the logs with oxen or horses over iced trails to the riverbank. There the logs were decked onto "rollways." In the spring when snow thawed and water levels rose, the logs were rolled into the river, and the drive commenced.

To ensure that logs drifted freely along the river, men called "log drivers" or "river pigs" were needed to guide the logs. The drivers typically divided into two groups. The more experienced and nimble men comprised the "jam" crew or "beat" crew. They watched the spots where logs were likely to jam, and when a jam started, tried to get to it quickly and dislodge the key logs before many logs stacked up. If they didn't, the river would keep piling on more logs, forming a partial dam which could raise the water level. Millions of board feet of lumber could back up for miles upriver, requiring weeks to break up, with some lumber possibly lost if it was shoved far enough into the shallows. So when the jam crew saw a jam start, they rushed to it and tried to break it up, using peaveys and possibly dynamite. This job required some understanding of physics, strong muscles, and extreme agility. The jam crew was an exceedingly dangerous occupation, with the drivers standing on the moving logs and running from one to another. Many drivers lost their lives by falling and being crushed by the logs.

Log driver working for the Brown Company in New Hampshire.

River drivers in Germany with a crude pike pole the Germans call a rafters' hook (Flößerhaken, as in rafting logs), probably one or two pickaroons and a ring dog for rolling logs over.

Each crew was accompanied by an experienced boss often selected for his fighting skills to control the strong and reckless men of his team. The overall drive was controlled by the "walking boss" who moved from place to place to coordinate the various teams to keep logs moving past problem spots. Stalling a drive near a saloon often created a cascade of drunken personnel problems.

A larger group of less experienced men brought up the rear, pushing along the straggler logs that were stuck on the banks and in trees. They spent more time wading in icy water than balancing on moving logs. They were called the "rear crew." Other men worked with them from the bank, pushing logs away with pike poles. Others worked with horses and oxen to pull in the logs that had strayed furthest out into the flats.

Bateaux ferried log drivers using pike poles to dislodge stranded logs while maneuvering with the log drive. A wannigan was a kitchen built on a raft which followed the drivers down the river. The wannigan served four meals a day to fuel the men working in cold water. It also provided tents and blankets for the night if no better accommodations were available. A commissary wagon carrying clothing, plug tobacco and patent medicines for purchase by the log drivers was also called a wangan. The logging company wangan train, called a *Mary Anne*, was a caravan of wagons pulled by four- or six-horse teams where roads followed the river to transport the tents, blankets, food, stoves, and tools needed by the log drivers.

For log drives, the ideal river would have been straight and uniform, with sharp banks and a predictable flow of water. Wild rivers were not that, so men cut away the fallen trees that would snag logs, dynamited troublesome rocks, and built up the banks in places. To control the flow of water, they built "flash dams" or "driving dams" on smaller streams, so they could release water to push the logs down when they wanted.

Each timber firm had its own mark which was placed on the logs, called an "end mark". Obliterating or altering a timber mark was a crime. At the mill the logs were captured by a log boom, and the logs were sorted for ownership before being sawn.

Log drives were often in conflict with navigation, as logs would sometimes fill the entire river and make boat travel dangerous or impossible.

Floating logs down a river worked well for the most desirable pine timber, because it floated well. But hardwoods were more dense, and didn't float well enough to be easily driven, and some pines weren't near drivable streams. Log driving became increasingly unnecessary with the development of railroads and the use of trucks on logging roads. However, the practice survived in some remote locations where such infrastructure did not exist. Most log driving in the United States and Canada ended with changes in environmental legislation in the 1970s. Some places, like the Catalonian Pyrenees, still retain the practice as a popular holiday celebration once a year.

Popular Culture

- The figure of speech "High and Dry" describes an unsuccessful log drive. Maximum river flows typically coincided with the runoff from snowmelt, and was sometimes augmented by water released from flash dams. If logs were started downriver when there was not enough water to move them all the way to the sawmill, the investment made in cutting that timber might be stranded *high and dry* in shallows along the stream for a year until the next spring snowmelt.
- The contemporary logrolling contest, Birling, is a demonstration of skills originally devised by log drivers.

- Inclusive description of a complex assortment as "the whole Mary Anne" derives from the colorful characters of wangan caravans which periodically transformed quiet rural communities with the excitement of a passing log drive.
- In Canada, "The Log Driver's Waltz" is a popular folk song about the practice.
- The version of the Canadian one-dollar bank note issued in 1974 and withdrawn in 1989 featured a view of the Ottawa River with log driving taking place in the foreground and Parliament Hill rising in the background. This banknote was part of the fourth series of banknotes released by the Bank of Canada entitled "Scenes of Canada". The logs depicted in this bank note may have been destined for a half dozen pulp, paper and sawmills near the Chaudière Falls immediately upstream from Parliament Hill, or for other mills further downstream.
- The song "Flottarkärlek" describes campfire stories from a log drivers perspective.
- An Englishman may have observed loggers loitering in Bangor, Maine when he reported in 1801: "His habits in the forest and the [river] voyage all break up the system of persevering industry and substitute one of alternate toil and indolence, hardship and debauch; and in the alteration, indolence and debauch will inevitably be indulged in the greatest possible proportion."

Log Scaler

The log scaler is an occupation in the timber industry. The Log Scaler measures the cut trees to determine the scale (volume) and quality (grade) of the wood to be used for manufacturing. When logs are sold, in order to determine the basis for a sale price in a standard way, the logs are "scaled" which means they are measured, identified as to species, and deductions for defects assigned to produce a net volume of merchantable wood. There are several different scales or rules that are used to determine the volume of wood. Scribner Decimal C rule is based on diagrams of circles that show the amount of boards that will be utilized from diameters of logs. Cubic rule, often called Metric in Canada, determines the cubic volume of the log material. The logs are recorded as gross scale (actual log measurements, length and diameter) and net scale (volume after deductions for defects are taken out). This occupation is usually performed by a third party organization qualified to "scale" government timber. Since internal defects are determined by external indications it is not an exact science and is subject to interpretation of log scaling rules. The log scaler is subject to random "check scales" in which another scaler rescales exactly the same logs and the results are compared. The log scaler must be within + or - 1% of the gross scale and + or - 2% of the net scale to keep their certification to scale. The scale is used for payment, quality control and inventory purposes.

Historical Methods

Historically in the Pacific Northwest logs were first sold after they were delivered to the water and rafted to the mill. Bundle rafts have bundles of logs that are tightly fastened together and some logs are well below the water line. Consequently, log scalers would walk the rafts and measure and grade the logs in the water. When logging was done by hand, the fallers and buckers were often paid on a piecework basis so the individual workmen's production was scaled daily as well. Presently few logs are ever rafted.

Ramp Scaling

Ramp Scale at Mist, Oregon

A log scaling method that has been popular since the adaptation of trucks to haul logs has been the 'ramp scale'. In this approach the log truck is driven beside a scale ramp which is simply an elevated platform and from there the log scaler climbs around on the truck and on the load and measures and grades the logs. Although the ramp is typically operated by the scaling bureau it may be located with a sawmill yard that is buying the logs or it may be at some other convenient location. Depending on the location the ramp may scale logs for a single mill, or the ramp may be located so that logs scaled may proceed to multiple destinations. The photo at the right shows two trucks at a typical Pacific Northwest bureau scale ramp. Ramp scaling is often used where space is limited, speed is important, or the logs are of relatively low value. Assessing the grade of a log involves inspecting the log for defects and when it is in the middle of a truck load of logs it may be difficult to see the log.

Rollout Scaling

Roll out scaling in Oregon Log yard

The preferred method of scaling, particularly when high value logs are involved, is the 'roll out' method. In this process the logs are unloaded from the truck and rolled out on the ground or on skids a single layer deep so the scaler can then see a good part of every log to assist in determining what if any defects it may have such as bend or rot. Roll out scaling has to be done where heavy machinery is available to handle the logs and requires the space to do the rollout as well as the time and expense of the extra handling. Rollout scaling can be more efficient (at least for the trucks) as the trucks do not have to wait to be scaled and can be unloaded and leave immediately, this contrasts with the ramp scale where the truck must wait until the scaling is complete. At times there are still places where the logs are coming in on a railroad and the car can be left for roll out scale. Timber Cruisers also estimate the volume of timber, for example in a timber sale where the timber is sold on the stump, and can see the logs and where limbs grew on a tree, especially the lower logs in a tree which have the most volume and often the best grade of log and thus often the most value, better than a scaler. However roll out scale helps the scaler do that better than truck scale or, up in British Columbia, in a log raft. The better the quality and bigger the size of logs are, the better it is to scale logs when they are "rolled out".

Standing Wood Scaling

Method of standing wood scaling is used for high-value logs volume measurement or for creation of allometric equations.

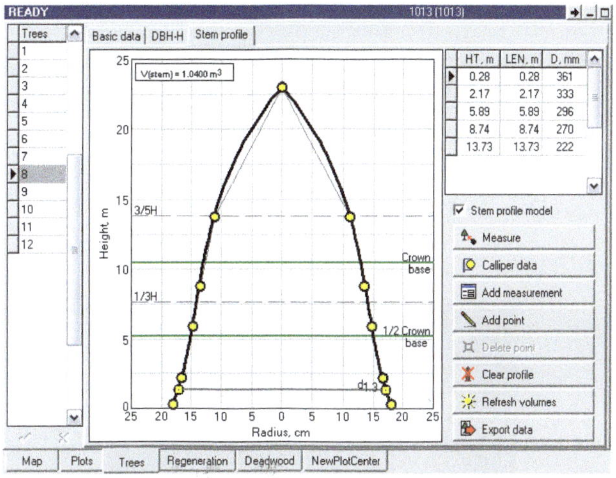

Stem profile (measured by special scope) allows to obtain accurate stem volume.

Rubber Tapping

Rubber tapping is the process by which the latex is collected from a rubber tree. An incision is made in the tree's bark, which cuts through the latex vessels, from which the product then flows. Timing of the incision must be planned within the planting cycle to optimise the latex yield.

Rubber tapping is an environmentally attractive land use. Jungle rubber is essentially old secondary forest, strongly resembling the primary forest. Its species' richness is about half that of the primary forest. Michon and de Foresta (1994) found that sample jungle rubber sites contained 92 tree species, 97 lianas, and 28 epiphytes compared to 171, 89, and 63, respectively, in the primary forest, and compared to 1, 1, and 2 in monoculture estates. Thiollay (1995) estimated that jungle rubber supports about 137 bird species, against 241 in the primary forest itself. Jungle rubber is expected to resemble primary forest in its hydrological functions. Monoculture rubber tree plantations have far less of an environmental impact than other crops, such as coffee or especially oil palm.

Rubber tapping Sri Lanka

Process

Rubber Tapping in Malaya, circa 1910

Rubber tapping in Basilan, Philippines.

Each night a rubber tapper must remove a thin layer of bark along a downward half spiral on the tree trunk. If done carefully and with skill, this tapping panel will yield latex for up to five years. Then the opposite side will be tapped, allowing this side to heal over. The spiral allows the latex to run down to a collecting cup. The work is done at night or in the early morning before the day's temperature rises, so the latex will drip longer before coagulating and sealing the cut.

Depending on the final product, additional chemicals can be added to the latex cup to preserve the latex longer. Ammonia solution helps prevent natural coagulation and allows the latex to remain in its liquid state. Plastic bags containing a coagulant have replaced cups in many plantations in Malaysia. This form of latex is used as the raw material for latex concentrate, which is used for dipped rubber products or for the manufacture of ribbed smoke sheet grades.

Naturally coagulated latex, sometimes referred to as cup lump, is collected for processing into block rubbers, which are referred to as technically specified rubbers (TSRs). The serum left after latex coagulation is rich in quebrachitol, a cyclitol or cyclic polyol.

In Information Technology

In information technology the term 'Rubber tapping' is sometimes used as a lampoon for planned Production deployments, where the timing of the deployment is crucial. Such deployments are generally done late in the night when customer activity is minimum.

For high availability information technology services such as data centres, telecommunications, Internet service providers etc., a time period is stated in advance to the customer to prepare for any possible service disruption to bring about any change to the service. Such a disclosure is typically guaranteed as part of a service level agreement.

A maintenance window is scheduled between 02:00 AM to 06:00 AM which is the possible low traffic and customer usage time frame. As such, the IT or network engineer who is responsible in such service change and/or upgrade activity are called "Rubber Tappers"; for which they are required to wake up early in the morning to complete their deployment, testing and commissioning before sunrise.

Tree Breeding

Tree breeding is the application of genetic, reproductive biology and economics principles to the genetic improvement and management of forest trees. In contrast to the selective breeding of livestock, arable crops, and horticultural flowers over the last few centuries, the breeding of trees, with the exception of fruit trees, is a relatively recent occurrence.

A typical forest tree breeding program starts with selection of superior phenotypes (plus trees) in a natural or planted forest. This application of mass selection improves the mean performance of the forest. Offspring is obtained from selected trees and grown in test plantations that act as genetic trials. Based on such tests the best genotypes among the parents can be selected. Selected trees are typically multiplied by either seeds or grafting and seed orchards are established when the preferred output is improved seed. Alternatively, the best genotypes can be directly propagated by cuttings or in-vitro methods and used directly in clonal plantations. The first system is frequently used in pines and other conifers, while the second is typical in some broadleaves (poplars, eucalypts and others). The objectives of a tree breeding program range from yield improvement and adaptation to particular conditions, to pest- and disease-resistance, wood properties, etc. Currently, tree breeding starting to take advantage of the fast development in plant genetics and genomics.

Optimisation

Tree breeders make efforts to get their operation efficient by optimising tree breeding. Scientists develop tools aimed at improvement of the efficiency of tree breeding programmes. Optimising can mean adapting strategies and methods to certain species, group of populations, structure of genetic variation and mode of inheritance of the important traits to obtain the highest benefit per unit of time. Optimising is usually carried out at the following levels: breeding strategy (appropriate intensity of breeding, breeding population structure and size, plan for maintenance of genetic diversity), breeding methods (mating type, testing and selection methods, testing population size and time) and deployment methods of the genetically improved material (seed orchards and clonal forestry: genetic contribution, size). Computer simulators are frequently used: stochastic – based on defined and random algorithms; and deterministic – based on defined algorithms.

Selection strategies have been compared for annual progress in long-term breeding at a given annual cost considering genetic gain, gene diversity, cost components, and time components. For Norway spruce it seems favourable to clone full sib families and then select based on clonal performance while for Scots pine a two-stage strategy seems best, first phenotypic pre-selection and then progeny-testing the selections.

Tree Improvement

A genetically variable population and a method of selecting genetically superior individuals provide the basis for tree improvement by breeding. In essence, a tree improvement program sets out to isolate and evaluate the genetic component of variation in one or more characters of interest. In the simplest procedure, cycles of selection reduce the available population in a particular direction to enhance desirable traits, then breeding from selections to expand the population with improved characteristics. Breeding strategies vary with species and objectives, but all use mating designs to generate information and new material. Choice of a suitable breeding strategy and mating design is a key decision in any breeding program. Kiss (1986) used a 2-level design in British Columbia to study variation within and between separate populations of white spruce, both within British Columbia and from eastern North America.

The breeding program for white spruce initiated in 1986 by the Canadian Forestry Service in the Maritimes employed 2 kinds of mating: polycross, to test clones for general combining ability; and pair-mating, to generate material for second generation selections (Fowler et al. 1988).

Newton's (2003) systematic review of yield responses of white spruce and 3 other North American conifers to forest tree improvement practices indicated that correct provenance-progeny selection could yield juvenile height growth gains of about 12% at 20 years for white spruce, and a corresponding merchantable productivity (mean annual merchantable volume increment) gain of 26% at 50 years for plantations established at nominal initial densities on medium-to-good quality sites. Also, preliminary estimates derived from individual case studies indicated that first generational selection strategies for white spruce could increase merchantable productivity by approximately 20% at 45 years.

Cable Logging

High Lead logging in Western Oregon

Cable grue *Larix 3T, installed on agricultural tractor*

Cable logging or skyline logging is a logging method primarily used on the West Coast of North America with yarder, loaders and grapple yarders, but also in Europe (Austria, Switzerland, Czech Republic, France, Italy).

The cables can be rigged in several configurations.

There are two classes.

1. High lead logging, in which a simple loop of cable runs from the yarder out through pulley blocks anchored to stumps at the far end of the cut.
2. Skyline, in which a carriage, pulled by hauling cables, runs along a skyline cable, providing vertical lift to the logs.

Into this there are varieties of loading systems.

While skyline logging requires additional setup time, the vertical lift of the skyline allows faster yarding, which can outweigh the additional labor costs, especially on larger harvest units.

Since the 1980s Grapple Yarders have become popular.

Skyline and grapple yarding, however, require more complex and expensive equipment. A traditional highlead or gravity system will function with just two cables, a skidding line and a haulback line. As the names imply the skidding line is used to drag the logs in, and the haulback line is used to drag the skidding line back out for the next turn (or group of logs).

A skyline system will add a third line---the skyline whose function is to hold the skidding line and the haulback line off the ground or 'in the sky'.

The yarder itself is located on a landing, a flat area on top of the ridge that is being logged. After the trees are retrieved by the yarder, the limbs are bucked (removed) and the logs are then placed in piles awaiting transport.

Clearcutting

After a century of clearcutting, this forest, near the source of the Lewis and Clark River in Clatsop County, Oregon, is a patchwork. In each patch, most of the trees are the same age.

Clearcutting, clearfelling, or clearcut logging is a forestry/logging practice in which most or all trees in an area are uniformly cut down. Clearcutting, along with shelterwood and seed tree harvests, is used by foresters to create certain types of forest ecosystems and to promote select species that require an abundance of sunlight or grow in large, even-age stands. Logging companies and forest-worker unions in some countries support the practice for scientific, safety, and economic reasons. Detractors see clearcutting as synonymous with deforestation, destroying natural habitats and contributing to climate change.

Clearcutting is the most common and economically profitable method of logging. However, clearcutting also imposes other externalities in the form of detrimental side effects such as loss of topsoil; the value of these costs is intensely debated by economic, environmental, and other interests. Aside from the purpose of harvesting wood, clearcutting is also used to create land for farming. The "insatiable human demand for wood and arable land" through clearcutting and other activities has led to the loss of over half of the world's rainforests.

While deforestation of both temperate and tropical rainforests through clearcutting has received considerable media attention in recent years, the other large forests of the world, such as the taiga, also known as boreal forests, are also under threat of rapid development. The same reasons for preserving the world's tropical rainforests also apply to the taiga, as do the reasons for destroying them. In Russia, as in North America and Scandinavia, creating protected areas and granting long-term leases to tend and regenerate trees—thus maximizing future harvests—are ways of limiting the harmful effects of clearcutting. Long-term studies of clearcut forests, such as studies of the Pasoh Rainforest in Malaysia, are also important in providing insights into the preservation of forest resources worldwide.

Types

Many variations of clearcutting exist; the most common professional practices are:

- *Standard (uniform) clearcut* – removal of every stem (whether commercially viable or not), so no canopy remains.
- *Patch clearcut* – removal of all the stems in a limited, predetermined area (patch).
- *Strip clearcut* – removal of all the stems in a row (strip), usually placed perpendicular to the prevailing winds in order to minimize the possibility of windthrow.
- *Clearcutting-with-reserves* – removal of the majority of standing stems save a few reserved for other purposes (for example as snags for wildlife habitat), (often confused with the seed tree method).
- *Slash-and-burn* – the permanent conversion of tropical and subtropicals forests for agricultural purposes. This is most prevalent in tropical and subtropical forests in overpopulated regions in developing and least developed countries. Slash-and-burn entails the removal of all stems in a particular area. This can be a form of deforestation, when the land is converted to other uses. However, some indigenous forest peoples, for example the 19th century Forest Finns rotate over the land and it does return to forest and this would be sustainable. Slash and burn techniques are typically used by civilians in search of land for living and agricultural purposes. The forest is first clear cut, and the remaining material is burned. One of the driving forces behind this process is a result of overpopulation and subsequent sprawl. These methods also occur as a result of commercial farming. The lumber is sold for profit, and the land, cleared of all remaining brush and suitable for agricultural development, is sold to farmers.
- *Selection cutting* – which can be done for timber harvesting or for ecological reasons. when so done it is often called ecoforestry.

Clearcutting contrasts with selective cutting, such as high grading, in which only commercially valuable trees are harvested, leaving all others. This practice can reduce the genetic viability of the forest over time, resulting in poorer or less vigorous offspring in the stand. Clearcutting also differs from a coppicing system, by allowing revegetation by seedlings. Additionally, destructive forms of forest management are commonly referred to as 'clearcutting'.

However it has been argued that in New Zealand clearcutting is more environmentally friendly as it uses less resources for logging and less land for wood production allowing other land to be kept as old growth forest.

Clearcutting Regeneration, Harvesting or System

Clearcutting can be differentiated into

- *Clearcutting* - clean felling by complete exploitation and removal of all the trees in one operation ... a harvesting method
- *Clearcutting method* - method for regenerating an even-aged community by removing all the mature trees
- *Clearcutting system* - a silvicultural system incorporating the clearcutting method to remove (clear) the mature community over a considerable area at one time

Confusion between these different uses of the term is common. Furthermore as indicated above many variations mean technically correct usage may not be descriptive enough to know what is meant in that particular occasion.

Clearcutting in Southern Finland

Clearcutting near Eugene, Oregon

Effects on the Environment

Environmental groups criticize clear-cutting as destructive to water, soil, wildlife, and atmosphere, and recommend the use of sustainable alternatives. Clear-cutting has a very big impact on the water cycle. Trees hold water and topsoil. Clear-cutting in forests removes the trees which would otherwise have been transpiring large volumes of water and also physically damages the grasses, mosses, lichens, and ferns populating the understorey. All this bio-mass normally retains water during rainfall. Removal or damage of the biota reduces the local capacity to retain water, which can exacerbate flooding and lead to increased leaching of nutrients from the soil. The maximum nutrient loss occurs around year two, and returns to pre-clearcutting levels by year four.

Clear-cutting also prevents trees from shading riverbanks, which raises the temperature of riverbanks and rivers, contributing to the extinction of some fish and amphibian species. Because the trees no longer hold down the soil, riverbanks increasingly erode as sediment into the water, creating excess nutrients which exacerbate the changes in the river and create problems miles away, in the sea. All of the extra sediment and nutrients that leach into the streams cause the acidity of the stream to increase, which can kill marine life if the increase is great enough. The nutrient content of the soil was found to return to five percent of pre-clearcutting levels after 64 years, which demonstrates how clearcutting affects the environment for many years.

Clearcutting can destroy an area's ecological integrity in a number of ways, including: the destruction of buffer zones which reduce the severity of flooding by absorbing and holding water; the immediate removal of forest canopy, which destroys the habitat for many rainforest-dependent insects and bacteria; the removal of forest carbon sinks, leading to global warming through the increased human-induced and natural carbon dioxide build-up in the atmosphere; the elimination of fish and wildlife species due to soil erosion and habitat loss; the removal of underground worms, fungi and bacteria that condition soil and protect plants growing in it from disease; the loss of small-scale economic opportunities, such as fruit-picking, sap extraction, and rubber tapping; and the destruction of aesthetic values and recreational opportunities.

Negative Impacts

Clearcutting can have major negative impacts, both for humans and local flora and fauna. A study from the University of Oregon found that in certain zones, areas that were clear cut had nearly three times the amount of erosion due to slides. When the roads required by the clearcutting were factored in, the increase in slide activity appeared to be about 5 times greater compared to nearby forested areas. The roads built for clearcutting interrupt normal surface drainage because the roads are not as permeable as the normal ground cover. The roads also change subsurface water movement due to the redistribution of soil and rock. Clearcutting may lead to increased stream flow during storms, loss of habitat and species diversity, opportunities for invasive and weedy species, and negative impacts on scenery, as well as a decrease in property values; diminished recreation, hunting, and fishing opportunities. Clearcutting decreases the occurrence of natural disturbances like forest fires and natural uprooting. Over time, this can deplete the local seed bank. An example of what clearcutting did in Ontario before 1900 can be found in Edmund Zavitz.

In temperate and boreal climates, clearcutting can have an effect on the depth of snow, which is usually greater in a clearcut area than in the forest, due to a lack of interception and evapotranspiration. This results in less soil frost, which in combination with higher levels of direct sunlight results in snowmelt occurring earlier in the spring and earlier peak runoff.

The world's rain forests could completely vanish in a hundred years at the current rate of deforestation. Between June 2000 and June 2008 more than 150 000 square kilometers of rain forest were cleared in the Brazilian Amazon. Huge areas of forest have already been lost. For example, only eight to fourteen percent of the Atlantic Forest in South America now remains. While deforestation rates have slowed since 2004, forest loss is expected to continue for the foreseeable future. Farmers slash and burn large parcels of forest every year to create grazing and crop lands, but the forest's nutrient-poor soil often renders the land ill-suited for agriculture, and within a year or two, the farmers move on.

Positive Perspectives

Clearcutting can be practiced to encourage the growth and proliferation of tree species that require high light intensity. Generally, a harvest area wider than double the height of the adjacent trees will no longer be subject to the moderating influence of the woodland on the microclimate. The width of the harvest area can thus determine which species will come to dominate. Those with high tolerance to extremes in temperature, soil moisture, and resistance to browsing may be established, in particular secondary successional pioneer species.

Clearcutting can be used by foresters as a method of mimicking a natural disturbance and increasing primary successional species, such as poplar (aspen), willow and black cherry in North America. Clearcutting has also proved to be effective in creating animal habitat and browsing areas, which otherwise would not exist without natural stand-replacing disturbances such as wildfires, large scale windthrow, or avalanches.

Clearcuts are used to help regenerate species that cannot compete in mature forests. A number of them are aspen, jack pine and, in areas with poor soils, oaks—are important species for both game and nongame wildlife species. Clearcutting can also lead to increased vascular-plant diversity in the area. This is most pronounced after a couple years of clearcutting and in herb-rich forests where scarification took place.

No significant changes in water temperature were observed when patch clearcutting was done 100 feet away from a river. This suggests that patch clearcutting is a possible solution to concerns about changes in water temperature due to clearcutting. The effects of clearcutting on soil nutrient content were not examined in this study.

More recently, forest managers have found that clearcutting oak stands helps regenerate oak forests in areas of poor soil. The tree canopies in oak forests often shade out the ground, making it impossible for newly sprouted oaks to grow. When the mature trees are removed, the saplings stand a chance of recruiting into the forest.

Effects on Wildlife

Clearcutting's main destruction is towards habitats, where it makes the habitats more vulnerable in the future to damage by insects, diseases, acid rain, and wind. Removal of all trees from an area destroys the physical habitats of many species in wildlife. Also clearcutting can contribute to problems for ecosystems that depend on forests, like the streams and rivers which run through them.

In Canada, the black-tailed deer population is at further risk after clearcutting. The deer are a food source for wolves and cougars, as well as First Nations and other hunters. While deer may not be at risk in cities and rural countryside, where they can be seen running through neighbourhoods and feeding on farms, in higher altitude areas they require forest shelter.

Tree Planting

A treeplanter in northern Ontario.

Tree planting is the process of transplanting tree seedlings, generally for forestry, land reclamation, or landscaping purpose. It differs from the transplantation of larger trees in arboriculture, and from the lower cost but slower and less reliable distribution of tree seeds.

Tree planting is an aspect of habitat conservation. In each plastic tube a hardwood tree has been planted.

In silviculture the activity is known as reforestation, or afforestation, depending on whether the area being planted has or has not recently been forested. It involves planting seedlings over an area of land where the forest has been harvested or damaged by fire, disease or human activity. Tree planting is carried out in many different parts of the world, and strategies may differ widely across nations and regions and among individual reforestation companies. Tree planting is grounded in forest science, and if performed properly can resulwt in the successful regeneration of a deforested area. Reforestation is the commercial logging industry's answer to the large-scale destruction of old growth forests, but a planted forest rarely replicates the biodiversity and complexity of a natural forest.

Because trees remove carbon dioxide from the air as they grow, tree planting can be used as a geoengineering technique to remove CO_2 from the atmosphere. Desert greening projects are also motivated by improved biodiversity and reclamation of natural water systems, but also improved economy and social welfare due to increased number of jobs in farming and forestry.

By Country

Australia

Australian forests have been heavily affected since European colonisation, and some attempts have been made to restore native habitats, both by government and individuals. Greening Australia is a national Non profit set up to run the "National Tree Program" initiated by the Federal Government in 1982. Greening Australia completed the 1 Billion Tree target and has gone on to become one of the major tree planting organisations in the country. There is a strong volunteer movement for conservation in Australia through Landcare and other networks. National Tree Day is organised annually by Planet Ark in the last week in July, encouraging the public to plant 1 million native trees per year.Growing trees for Timber industries is a long term project. It may take many years for a tree to mature to an age and size that is appropriate for the Timber to be used by industry. Some trees are many hundreds of years old.

Many state governments run their own "Million Tree" programs each year to encourage community involvement.

Canada

Most tree planting in Canada is carried out by private reforestation companies. Reforestation companies compete with one another for contracts from logging companies, whose annual allowable cut for the following year is based upon how much money they invest into reforestation and other silvicultural practices. Treeplanting is typically piece work and tree prices can vary widely depending on the difficulty of the terrain and on the winning contract's bid price. As a result, there is a saying among planters: "There is no bad land, only bad contracts." 4 months of hard work can yield enough to live on for an entire year, but conditions are harsh.

Tree planting crews often do not permanently reside in the areas where they work, thus much planting is based out of motels or bush camps. Bush camp accommodations usually consist of a mess tent, cook shack, dry goods tent, first aid tent, freshly dug outhouses, and a shower tent or trailer. Planters are responsible for bringing either a tent or car to sleep in. A camp also contains camp cooks and support staff.

Planting is carried out in accordance to the client's specifications, and planters are expected to learn the quality standards for each contract that they work on. Planted clearcuts are spot checked on a regular basis. Although quality concerns vary across contracts, spot checkers are typically looking for such things as: species appropriate site choice, species appropriate spacing, how tight the seedlings are in the ground, how straight the seedlings are, and whether or not the seedlings have been damaged. These concerns vary from region to region, and from contract to contract.

The average British Columbian planter plants 1 600 trees per day, but it is not uncommon for veterans to plant up to 4,000 trees per day while working in the interior. These numbers are higher in central and eastern Canada, where the terrain is generally faster, however the price per tree is slightly lower as a result. Average daily totals of 2500 are common, with experienced planters planting upwards of 5000 trees a day. Numbers as high as 7500 a day have been recorded. Planters typically work 8–10 hours per day with an additional 1 to 2 hours of (usually) unpaid traveling time. Work weeks on British Columbian planting contracts are usually 4–5 days long, with 1–2 days off. In Ontario, work weeks are generally 5–6 days long, with 1 day off.

Quite often, tree planting contractors will deduct some of the cost associated with the operation of the contract directly from the tree planter's daily earned wages. These imposed fees typically vary from $10 to $30 per day, and are referred to as "camp costs". In *some* cases, rookie tree planters end up owing their employer money for the first few pay periods.

Once inflation is factored in, real tree planter earnings have declined for many years in Canada. This has adversely affected the sector's ability to attract and retain workers. Higher wages and much better working conditions in many other industries, from construction, to oil and gas, and even information technology, has led to fewer Canadian young people wanting to plant trees.

Based on statistics for British Columbia, the average tree planter: lifts a cumulative weight of over 1,000 kilograms (2,200 lb), bends more than 200 times per hour, drives the shovel into the ground more than 200 times per hour and travels over 16 kilometres (9.9 mi) with a heavy load, every day of the entire season. The reforestation industry has an average annual injury rate of approximately 22 claims per 100 workers, per year. It is often difficult and sometimes dangerous.

Great Britain

Planting in Britain is commonly referred to as *restocking*, when it takes place on land that has recently been harvested. When occurring on previously unforested land it is known as *new planting*. Under the British system, in order to acquire the necessary permissions to clearcut, the landowner must agree a management plan with the Forestry Commission (the regulatory body for all things forestry) which must include proposals for the re-establishment of tree cover on the land. Planting contractors will be engaged by the landowner/management company, a contract drawn up and work will typically take place from November to April when most of the transplants are dormant.

Planting is part of the rotational nature of much British plantation forestry. Productive tree crops are planted and subsequently clearcut. Some form of soil cultivation may take place and the ground is then restocked. Where the production of timber is a management priority, a prescribed stocking density must be achieved. For coniferous species this will be a minimum of 2500 stems per hectare at year 5 (from planting). Planting at this density has been shown to favour the development of straighter knot-free logs.

Planters are normally paid under piece work terms and an experienced worker will plant around 1500 trees a day under most conditions.

Israel

Tree-planting is an ancient Jewish tradition. The Talmudic rabbi Yohanan ben Zakai used to say that if a person planting a tree heard that the Messiah had arrived, he should finish planting before going to greet him. With over 240 million planted trees, Israel is one of only two countries that entered the 21st century with a net gain in the number of trees. Due to massive afforestation efforts, this fact echoed in diverse campaigns. Israeli forests are the product of a major afforestation campaign by the Jewish National Fund (JNF).

The largest planted forest in Israel is Yatir Forest, located on the southern slopes of Mount Hebron, on the edge of the Negev Desert. It covers an area of 30,000 dunams (30 square kilometers). It is named after the ancient Levite city within its territory, Yatir, as written in the Torah: "And unto the children of Aaron the priest they gave Hebron with its suburbs, the city of refuge for the manslayer, and Libnah with its suburbs, and Jattir with its suburbs, and Eshtemoa with its suburbs" (Book of Joshua 21:13-14). In 2006, the JNF signed a 49-year lease agreement with the State of Israel which gives it control over 30,000 hectares of Negev land for the development of forests. Research on climate change is being carried out in Yatir Forest. Studies of the Weizmann Institute of Science, in collaboration with the Desert Research Institute at Sde Boker, have shown that the trees function as a trap for carbon in the air. Shade provided by trees planted in the desert also reduces evaporation of the sparse rainfall. Yatir Forest is a part of the NASA project FluxNet, a global network of micrometeorological tower sites used to measure the exchanges of carbon dioxide, water vapor, and energy between terrestrial ecosystem and atmosphere. The Arava Institute for Environmental Studies conducts research that focuses on crops such as dates and grapes grown in the vicinity of Yatir forest. The research is part of a project aimed at introducing new crops into arid and saline zones.

The JNF has been criticized for planting non-native pine trees which are unsuited to the climate, rather than local species such as olive trees. Others say that JNF deserves credit for this decision, and the forests would not have survived otherwise. According to JNF statistics, six out of every 10 saplings planted at a JNF site in Jerusalem do not survive, although the survival rate for planting sites outside Jerusalem is much higher – close to 95 percent. Critics argue that many JNF lands outside the West Bank were illegally confiscated from Palestinian refugees, and that the JNF furthermore should not be involved with lands in the West Bank. Shaul Ephraim Cohen has claimed that trees have been planted to restrict Bedouin herding. Susan Nathan wrote that forests were planted on the site of abandoned Arab villages after the 1948 war. Nathan also writes that olive trees were replaced by pine and cypress trees and that JNF afforestation policy erases traces of the Arab presence prior to 1948.

Since 2009, the JNF has provided the Palestinian Authority with 3,000 tree seedlings for a forested area being developed on the edge of the new city of Rawabi, north of Ramallah.

Approximately one thousand small forest fires are registered on average every year during the five fire-prone months. Half of them are caused by arson, hostile actions and Arab or Palestinian terrorist attacks. Ten thousand acres of hand-planted forest were destroyed by Katyusha rockets during the 2006 Lebanon War by Hezbollah. In summer 2006, JNF launched Operation Northern Renewal, a reforestation effort, which also replaced some topsoil that was burned away.

New Zealand

Tree planting by a boy in India

Kaingaroa Forest in New Zealand is the largest planted forest in the southern hemisphere. It is one of the many plantation forests planted since European settlement. The Monterey Pine (*Pinus radiata*) is commonly used for plantations since a fast-growing cultivar suitable for a wide range of conditions has been developed.

Government agencies, environmental organisations and private trusts carry out tree planting for conservation and climate change mitigation. While some work is carried out by private enterprise, there are also planting days organised for volunteers. Landcare Research use planted forests for their EBEX21 system for greenhouse gas emissions mitigations.

South Africa

South Africa's forests have been heavily depleted mostly due to agriculture, traditional farming and urbanisation in the coastal regions. Various organizations are working on increasing the forest cover in parts of the country. Currently there is less than 0.5% forest cover in South Africa. Greenpop is a national Social Enterprise set up to run a "Tree Planting Program" greening both urban and rural areas. This was initiated in 2010. There is a strong volunteer movement for conservation in South Africa. National Tree Day or Arbor Day is organised annually in September, and has gone on to become national Arbor Month.

United States of America

Trees for the Future and Plant With Purpose are non-profit organizations based in the U.S. that plant trees in developing countries to improve land management. Other organizations that plant trees in the United States include:

- American Forests
- Arbor Day Foundation
- Nature Conservancy
- Plant-it 2020
- USDA Forest Service "Plant-A-Tree" program in which a person can donate to plant trees in the National Forests.
- Our City Forest
- TreeFolks empowers central Texans to build stronger communities through planting and caring for trees. Since 1989, TreeFolks has planted over 1.5 million trees in parks, neighborhoods, and natural areas throughout central Texas.

Role in Climate Change

The development of markets for tradeable pollution permits in recent years have opened up a new source of funding for tree planting projects: carbon offsets. The creation of carbon offsets from tree planting projects hinges on the notion that trees help to mitigate climate change by sequestering carbon dioxide as they grow. However, the science linking trees and climate change is largely unsettled, and trees remain a controversial source of offsets.

Climate Impacts

Climate scientists working for the IPCC believe human-induced global deforestation is responsible for 18-25% of global climate change. The United Nations, World Bank and other leading nongovernmental organizations are encouraging tree planting to mitigate the effects of climate change.

Trees sequester carbon through photosynthesis, converting carbon dioxide and water into molecular dioxygen (O_2) and plant organic matter, such as carbohydrates (e.g., cellulose). Hence, forests that grow in area or density and thus increase in organic biomass will reduce atmospheric CO_2 levels. (Carbon is released as CO_2 if a tree or its lumber burns or decays, but as long as the forest is able to grow back at the same rate as its biomass is lost due to oxidation of organic carbon, the net result is carbon neutral.) In their 2001 assessment, the IPCC estimated the potential of biological mitigation options (mainly tree planting) is on the order of 100 Gigatonnes of carbon (cumulative) by 2050, equivalent to about 10% to 20% of projected fossil fuel emissions during that period.

However, the global cooling effect of forests from carbon sequestration is not the only factor to be considered. For example, the planting of new forests may initially release some of the area's existing carbon stores into the atmosphere. Specifically, the conversion of peat bogs into oil palm plantations has made Indonesia the world's third largest producer of greenhouse gases.

Compared to less vegetated lands, forests affect climate in three main ways:

- Cooling the Earth by functioning as carbon sinks, and adding water vapor to the atmosphere and thereby increasing cloudiness.
- Warming the Earth by absorbing a high percentage of sunlight due to the low reflectivity of a forest's dark surfaces. This warming effect, or reduced albedo, is large where evergreen forests, which have very low reflectivity, shade snow cover, which is highly reflective.

To date, most tree planting offsets strategies have taken only the first effect into account. A study published in December 2005 combined all these effects and found that tropical forestation has a large net cooling effect, because of increased cloudiness and because of high tropical growth and carbon sequestration rates.

Trees grow three times faster in the tropics than in temperate zones; each tree in the rainy tropics removes about 22 kilograms (50 pounds) of carbon dioxide from the atmosphere each year. However, this study found little to no net global cooling from tree planting in temperate climates, where warming due to sunlight absorption by trees counteracts the global cooling effect of carbon sequestration. Furthermore, this study confirmed earlier findings that reforestation of colder regions — where long periods of snow cover, evergreen trees, and slow sequestration rates prevail — probably results in global warming. According to Ken Caldeira, a study co-author from the Carnegie Institution for Science, "To plant forests outside of the tropics to mitigate climate change is a waste of time.".

His premise that grassland reflects more sun, keeping temperatures lower, is, however, applicable only in arid regions. A well-watered lawn, for example, is as green as a tree, but absorbs far less CO_2. Deciduous trees also have the advantage of providing shade in the summer and sunlight in the winter; so these trees, when planted close to houses, can be utilized to help increase energy efficiency of these houses.

This study remains controversial and criticized for assuming dark colored trees might replace the frozen, white tundra in the upper northern hemisphere. Regular tree planting projects typically take place on lands that are only slightly different in color. The warming impact was also measured over hundreds of years, rather than a 30-70 year time horizon most climate experts believe we have to fix climate change.

Furthermore, the described warming effect (of temperate and boreal latitude forest) is only apparent once the trees have grown to create a dense 'close canopy', and it is at precisely this point that trees grown for offset purposes should be harvested and their absorbed carbon fixed for the long-term as timber.

Costs

While the benefits of tree planting are subject to debate, the costs are low compared to many other mitigation options. The IPCC has concluded that "The mitigation costs through forestry can be quite modest (US$0.1–US$20 / metric ton carbon dioxide) in some tropical developing countries.... The costs of biological mitigation, therefore, are low compared to those of many other alternative measures". The cost effectiveness of tropical reforestation is due not only to growth rate, but also to farmers from tropical developing countries who voluntarily plant and nurture tree species which can improve the productivity of their lands. As little as US$90 will plant 900 trees, enough to annually remove as much carbon dioxide as is annually generated by the fossil-fuel usage of an average United States resident.

Types of Trees Planted

A eucalyptus plantation in final stages at Arimalam.

The type of tree planted may have great influence on the environmental outcomes. It is often much more profitable to outside interests to plant fast-growing species, such as eucalyptus, casuarina or pine (e.g., *Pinus radiata* or *Pinus caribaea*), even though the environmental and biodiversity benefits of such monoculture plantations are not comparable to native forest, and such offset projects are frequently objects of controversy.

To promote the growth of native ecosystems, many environmentalists advocate only indigenous trees be planted. A practical solution is to plant tough, fast-growing native tree species which begin rebuilding the land. Planting non-invasive trees that assist in the natural return of indigenous species is called "assisted natural regeneration." There are many such species that can be planted, of which about 12 are in widespread use, such as *Leucaena leucocephala*. Alternatively, farmer-managed natural regeneration (FMNR), involves farmers preserving trees (not replanting), and is considered to be a more cost effective method of reforestation than regular tree planting.

Season of Planting

Planting trees in USA

Bareroot Stock

The classical silvicultural literature unanimously advocates spring as the time to plant bareroot stock, with lifting and outplanting taking place while the trees are still apparently dormant. This view, in which spring planting is implicit, was epitomized by Toumey and Korstian (1942): "Almost without exception the most favourable time for ... planting is 2 weeks or more before buds [of the planting stock] begin their growth". Soil moisture conditions are generally favourable at the time when the growing season is about to begin, while dormant stock is less subject to mechanical injury and physiological shock.

If the size of the planting program allows, there is little doubt that such scheduling would be advantageous in that it satisfies one, and commonly 2, of the factors essential for success: (1) the use of planting stock that is physiologically capable of responding to a growth environment at planting, and (2) planting when site factors favour tree survival and growth. The 3rd factor a good planting job, and although desirable in all plantings, is probably somewhat less critical in conventional spring plantings than at other times. If, however, a planting program cannot be completed in this way, there are other options: conventional fall planting with fresh-lifted stock; summer planting with fresh-lifted stock; and spring and summer planting with stored spring-lifted or fall-lifted stock.

Conventional Spring Planting with Fresh Spring-lifted Stock

In the context of regeneration silviculture, "spring", "summer", etc. lack precise meaning. Typically, the spring planting season begins as soon as lifting becomes possible in the nursery, and ends with the completion of the program. At this time, planting stock is physiologically attuned to the oncoming growing season, and the outplant has the whole of that season in which to establish its root system before it is challenged by any frost heaving. In practice, ideals are seldom attained. That stock is normally dormant when spring-planted is a widespread fallacy. Active growth is commonly obvious at the time of planting, but in any case the metabolic activity increases in planting stock before the tops give visible expression to this. The difficulty of obtaining, in quantity, spring-lifted stock in dormant condition increases with increasing continentality of climate. In many areas, the period of spring-like weather is unreliable and often short. As well, the soil moisture advantage claimed for spring planting is also insecurely founded. Soils that are sandy or gravelly, and shallow soils of any texture are highly dependent on current weather due to their limited available water capacities. Nor will a plentiful supply of soil moisture benefit an outplant whose roots are enveloped in anaerobic and/or cold soil, and mortality of trees outplanted into soil colder than about 6 °C may be excessive. Flushing increases the vulnerability of planting stock to both physiological stress (because of increased water requirements and reduced root growth capacity) and mechanical damage, which probably accounts for the commonly asserted superiority of early spring over late spring planting.

In fact, the spruces may be planted not only throughout the spring planting period provided that the period of most active shoot elongation is avoided, but virtually throughout the whole growing season, with little loss of performance other than some reduction in increment. Mullin's (1971) study at Midhurst in southern Ontario illustrates both the success with which white spruce can be planted throughout the period in question and the need to minimize stresses on the planting stock. Mullin used 3+0 stock from regular shipping beds in a series of 6 weekly plantings beginning with apparently dormant trees on 3 May and ending on 7 June, by which time the new leading shoots were several centimetres long. Trees were lifted with and without root dipping, planted on the day of lifting after their root systems had been exposed for 0, 1, 2, or 3 hours. Whereas 2nd year survival in the control (root exposure = 0) condition varied little among the 6 plantings, with averages of 83.5% +/- 4.7% for root-dipped trees and 77.2 +/- 7.0% for non-dipped, mortality rates among root-exposed stock were very much more variable. For instance, 2nd-year survival among root-dipped trees whose root systems had been exposed for 1 hour varied from 17% to 84%.

Conventional Fall Planting with Fresh-lifted Stock

The fall planting season is generally considered to begin when nursery stock has hardened off and soil moisture reserves have been replenished by autumnal rain.It then continues until the planting program has been completed or is terminated by freeze-up or heavy snow. The advantages of fall planting were once considered "To outweigh those of spring so certainly" that in the National Forests of the Lake States almost all planting was done in the fall, but in spite of some success, operational fall plantings in North America have tended to be less successful than operational spring plantings. On certain sites, a major disadvantage of fall planting is that the root systems of outplants have little time in which to become firmly anchored before being subjected to frost heaving. Such plants are also vulnerable to "winter browning", which in fact may occur in the fall soon after planting, especially among stock having high shoot:root ratios. Relationships between dormancy progression and physiological condition, including root-growth capacity, are much less clear in the spruces than in the pines, but certainly there is good evidence that, in the absence of frost heaving, plantings of spruces can be just as successful in fall as in spring.

Summer Planting with Fresh-lifted Stock

Conceptually and logistically, the simplest way of extending the conventional planting season is to extend lifting and planting of fresh stock through summer until the planting program has been competed. There is ample evidence that spruces may be planted successfully throughout the summer, not least the notable operational success of Marek (Personal communication, 1985, George T. Marek, R.P.F., Beardmore, Ontario, Ontario Ministry of Natural Resources) in northwestern Ontario: "the success of my plantations is primarily due to different strategies and cool storaging in Beardmore. I have planted all species from spring to late fall, throughout the summer on a large scale, without any difficulties". Summer planting has also been successful in a number of research studies with white spruce, e.g., Crossley 1956; Ackerman and Johnson 1962; Decie 1962 cited by Revel and Coates 1976; Burgar and Lyon 1968; Mullin 1971, 1974; Revel and Coates 1976. Success depends on minimizing stresses to planting stock at all stages from lifting through planting and on planting when site conditions are conducive to survival and growth.

Spring and Summer Planting with Stored Stock

Refrigerated storage of planting stock has been developed largely with the aim of overcoming problems experienced in using flushed planting stock. Storage provides a means of holding stock for use when fresh stock is either unavailable or at a stage of development that renders it unsuitable for planting. It also offers possibilities of manipulating the physiological condition of the stock. However, there are problems associated with storage, e.g., mold, cold injury, desiccation, and depletion of food reserves. The rate of deterioration depends very much on the physiological condition of the planting stock at the time of lifting, as well as on the storage environment and duration of storage. In attempts to devise safe schedules for spring-lifting of stock for frozen storage, Mullin (1978) used a base of 0 °C and accumulated daily maximum soil temperatures at 15 cm depth to calculate degree days (DD). He interpreted the evidence to mean that white spruce destined for frozen storage should have accumulated no more than 50 DD before being lifted. With regard to cool-stored, spring-lifted stock, the main ingredients for success are lifting before flushing has begun, prevention of desiccation, maintenance of a constant temperature within 1 or 2 degrees of freezing, minimization of mold by good temperature control and sanitation, avoidance of crushing and other mechanical damage, and avoidance of longer than necessary periods of storage.

Mullin and Forcier (1976) and Mullin and Reffle (1980) examined the effects of spring-lifting date and planting date on several species, including 3+0 white spruce after frozen storage, with fresh-lifted controls planted on each planting date for comparison. In all plantings, the earliest (2 May) lifting gave highest average second-year survival in all species. In another study, Mullin (1978) found that outplantings of frozen-stored 3+0 white spruce were consistently successful to the end of July only with the earliest -lifted (25 April) stock. Sutton (1982) also used 3+0 white spruce in outplanting every 2 weeks from the end of June through the growing season in 3 successive years on a variety of sites in northern Ontario. Despite variation in planting stock, poor storage environments and adverse weather, 4th-year results showed a consistent pattern of reasonable survival and growth rates among trees planted through July, with a rapid decline in performance of trees planted thereafter. Overwinter storage of stock has also been employed. It has the advantage of lifting stock at the end of the growing season when physiological processes are invoking natural dormancy. Time of fall lifting was investigated by Mullin and Parker (1976) along with overwinter storage temperature to determine their effects on the performance of spring-planted 3+0 white spruce. There were 5 lifts, weekly from 19 October through 16 November, after which frozen ground put a stop to lifting. Two storage temperatures were used, -18 °C and -4 °C. Nearly all of the trees stored at -18 °C died. The other stock was planted in shallow furrows in sparsely sodded field of loamy sand on 12 April, 17 May, and 14 June along with fresh-lifted stock on each date. Fresh and stored white spruce gave comparable results in plantings extended into mid-June in the Midhurst area of Ontario.

Natural refrigerated overwinter storage has been used in root cellars and snow caches. Using natural refrigeration in root cellar storage, Jorgensen and Stanek (1962) kept 3+0 and 2+2 white spruce in dormant condition for 6 months without apparent detriment to performance after outplanting. Moreover, the stock was highly resistant to spring frost damage. Natural cold storage for overwintering 3+0 and 2+2 white spruce was also used by Mullin (1966). Unlike Jorgensen and Stanek's (1962) stock, which was raised 550 km to the south of where it was planted, Mullin's stock was raised in a nursery at about the same latitude as the planting site; the stock experienced inside-bale temperatures down to -15 °C in mid-winter, but still showed first- and second-year survival rates of 85.9% and 65.9%, respectively, compared with 91.4% and 76.2%, respectively, for fresh-lifted stock. However, Mullin's stored stock was much more damaged by spring frost than was fresh-lifted stock and it "showed a reduction in vigour as measured in terms of survival, susceptibility to damage and growth".

Plant Nursery

Plants in a nursery

A nursery is a place where plants are propagated and grown to usable size. They include retail nurseries which sell to the general public, wholesale nurseries which sell only to businesses such as other nurseries and to commercial gardeners, and private nurseries which supply the needs of institutions or private estates. Some retail and wholesale nurseries sell by mail.

Nurseries may supply plants for gardens, for agriculture, for forestry and for conservation biology.

Some nurseries specialize in one phase of the process: propagation, growing out, or retail sale; or in one type of plant: e.g., groundcovers, shade plants, or rock garden plants. Some produce bulk stock, whether seedlings or grafted, of particular varieties for purposes such as fruit trees for orchards, or timber trees for forestry. Some produce stock seasonally, ready in springtime for export to colder regions where propagation could not have been started so early, or to regions where seasonal pests prevent profitable growing early in the season.

Methods

An Orchid nursery

A tree nursery using gutters to decrease growing costs

Nurseries can grow plants in open fields, on container fields and in tunnels or greenhouses. In open fields nurseries grow ornamental trees, shrubs and herbaceous perennials, especially the plants meant for the wholesale trade or for amenity plantings. On a containerfield nurseries grow small trees, shrubs and herbaceous plants, usually destined for sales in garden centers. Nurseries also grow plants in greenhouses, a building of glass or in plastic tunnels, designed to protect young plants from harsh weather (especially frost), while allowing access to light and ventilation. Modern greenhouses allow automated control of temperature, ventilation and light and semi-automated watering and feeding. Some also have fold-back roofs to allow "hardening-off" of plants without the need for manual transfer to outdoor beds.

Most nurseries remain highly labor-intensive. Although some processes have been mechanised and automated, others have not. It remains highly unlikely that all plants treated in the same way at the same time will arrive at the same condition together, so plant care requires observation, judgment and manual dexterity; selection for sale requires comparison and judgment. A UK nurseryman has estimated (in 2003) that manpower accounts for 70% of his production costs. The largest UK nurseries have moved to minimize labour costs by the use of computer controlled warehousing methods: plants are pallet allocated to a location and grown on there with little human intervention. Picking merely requires selection of a batch and manual quality control before dispatch. In other cases, a high loss rate during maturation is accepted for the reduction in detailed plant maintenance costs.

Business is highly seasonal, concentrated in spring and autumn. There is no guarantee that there will be demand for the product - this will be affected by temperature, drought, cheaper foreign competition, fashion, among other things.

Annuals are sold in trays (undivided containers with multiple plants), flats (trays with built-in cells), peat pots, or plastic pots. Perennials and woody plants are sold either in pots, bare-root or balled and burlapped and in a variety of sizes, from liners to mature trees. Balled and Burlap (B & B) trees are dug either by hand or by a loader that has a tree spade attachment on the front of the machine. Although container grown woody plants are becoming more and more popular due to the versatility, B & B is still widely used throughout the industry.

Plants may be propagated by seeds, but often desirable cultivars are propagated asexually. The most common method is by cuttings. These can be taken from shoot tips or parts of stems with a node (softwood cuttings) or from older stems (hardwood cuttings). Herbaceous perennials are also often propagated by root cuttings or division. For plants on a rootstock grafting or budding is used. Older techniques like layering are sometimes used for crops which are difficult to propagate.

Conditioning

With the objective of fitting planting stock more ably to withstand stresses after outplanting, various nursery treatments have been attempted or developed and applied to nursery stock. Buse and Day (1989), for instance, studied the effect of conditioning of white spruce and black spruce transplants on their morphology, physiology, and subsequent performance after outplanting. Root pruning, wrenching, and fertilization with potassium at 375 kg/ha were the treatments applied. Root pruning and wrenching modified stock in the nursery by decreasing height, root collar diameter, shoot:root ratio, and bud size, but did not improve survival or growth after planting. Fertilization reduced root growth in black spruce but not of white spruce.

Hardening off, Frost Hardiness

Seedlings vary in their susceptibility to injury from frost. Damage can be catastrophic if "unhardenend" seedlings are exposed to frost. Frost hardiness may be defined as the minimum temperature at which a certain percentage of a random seedling population will survive or will sustain a given level of damage (Siminovitch 1963, Timmis and Worrall 1975). The term LT_{50} (lethal temperature for 50% of a population) is commonly used. Determination of frost hardiness in Ontario is based on electrolyte leakage from mainstem terminal tips 2 cm to 3 cm long in weekly samplings (Colombo and Hickie 1987). The tips are frozen then thawed, immersed in distilled water, the electrical conductivity of which depends on the degree to which cell membranes have been ruptured by freezing releasing electrolyte. A -15 °C frost hardiness level has been used to determine the readiness of container stock to be moved outside from the greenhouse, and -40 °C has been the level determining readiness for frozen storage (Colombo 1997).

In an earlier technique, potted seedlings were placed in a freezer chest and cooled to some level for some specific duration; a few days after removal, seedlings were assessed for damage using various criteria, including odour, general visual appearance, and examination of cambial tissue (Ritchie 1982).

Stock for fall planting must be properly hardened-off. Conifer seedlings are considered to be hardened off when the terminal buds have formed and the stem and root tissues have ceased growth. Other characteristics that in some species indicate dormancy are color and stiffness of the needles, but these are not apparent in white spruce.

Forest Tree Nurseries

Whether in the forest or in the nursery, seedling growth is fundamentally influenced by soil fertility, but nursery soil fertility is readily amenable to amelioration, much more so than is forest soil.

Nitrogen, phosphorus, and potassium are regularly supplied as fertilizers, and calcium and magnesium are supplied occasionally. Applications of fertilizer nitrogen do not build up in the soil to develop any appreciable storehouse of available nitrogen for future crops. Phosphorus and potassium, however, can be accumulated as a storehouse available for extended periods.

Fertilization permits seedling growth to continue longer through the growing season than unfertilized stock; fertilized white spruce attained twice the height of unfertilized. High fertility in the rooting medium favours shoot growth over root growth, and can produce top-heavy seedlings ill suited to the rigors of the outplant site. Nutrients in oversupply can reduce growth or the uptake of other nutrients. As well, an excess of nutrient ions can prolong or weaken growth to interfere with the necessary development of dormancy and hardening of tissues in time to withstand winter weather.

Stock Types, Sizes and Lots

Nursery stock size typically follows the normal curve when lifted for planting stock. The runts at the lower end of the scale are usually culled to an arbitrary limit, but, especially among bareroot stock, the range in size is commonly considerable. Dobbs (1976) and McMinn (1985a) examined how the performance of 2+0 bareroot white spruce related to differences in initial size of planting stock. The stock was regraded into large, medium, and small fractions according to fresh weight. The small fraction (20% of the original stock) had barely one-quarter of the dry matter mass of the large fraction at the time of outplanting. Ten years later, in the blade-scarified site, seedlings of the large fraction had almost 50% greater stem volume than had seedlings of the small fraction. Without site preparation, large stock were more than twice the size of small stock after 10 years.

Similar results were obtained with regraded 2+1 transplants sampled to determine root growth capacity. The large stock had higher RGC as well as greater mass than the small stock fraction.

The value of large size at the time of planting is especially apparent when outplants face strong competition from other vegetation, although high initial mass does not guarantee success. That the growth potential of planting stock depends on much more than size seems clear from the indifferent success of the transplanting of small 2+0 seedlings for use as 2+1 "reclaim" transplants. The size of bareroot white spruce seedlings and transplants also had a major influence on field performance.

The field performance among various stock types in Ontario plantations was examined by Paterson and Hutchison (1989): the white spruce stock types were 2+0, 1.5+0.5, 1.5+1.5, and 3+0. The nursery stock was grown at Midhurst Forest Tree Nursery, and carefully handled through lifting on 3 lift dates, packing, and hot-planting into cultivated weed-free loam. After 7 years, overall survival was 97%, with no significant differences in survival among stock types. The 1.5+1.5 stock with a mean height of 234 cm was significantly taller by 18% to 25% than the other stock types. The 1.5+1.5 stock also had significantly greater dbh than the other stock types by 30-43%. The best stock type was 57 cm taller and 1 cm greater in dbh than the poorest. Lifting date had no significant effect on growth or survival.

High elevation sites in British Columbia's southern mountains are characterized by a short growing season, low air and soil temperatures, severe winters, and deep snow. The survival and growth of Engelmann spruce and subalpine fir outplanted in 3 silvicultural trials on such sites in gaps of various sizes were compared by Lajzerowicz et al. (2006). Survival after 5 or 6 years decreased with smaller gaps. Height and diameter also decreased with decreasing size of gap; mean heights were 50 cm to 78 cm after 6 years, in line with height expectations for Engelmann spruce in a high-elevation planting study in southeastern British Columbia. In the larger gaps (≥1.0 ha), height increment by year 6 was ranging from 10 cm to 20 cm. Lajzerrowicz et al. Concluded that plantings of conifers in clearcuts at high elevations in the southern mountains of British Columbia are likely to be successful, even close to timberline; and group selection silvicultural systems based on gaps 0.1 ha or larger are also likely to succeed. Gaps smaller than 0.1 ha do not provide suitable conditions for obtaining adequate survival or for growth of outplanted conifers.

Planting Stock

Planting stock, "seedlings, transplants, cuttings, and occasionally wildings, for use in planting out," is nursery stock that has been made ready for outplanting.The amount of seed used in white spruce seedling production and direct seeding varies with method.

A working definition of planting stock quality was accepted at the 1979 IUFRO *Workshop on Techniques for Evaluating Planting Stock Quality* in New Zealand: "The quality of planting stock is the degree to which that stock realizes the objectives of management (to the end of the rotation or achievement of specified sought benefits) at minimum cost. Quality is fitness for purpose." Clear expression of objectives is therefore prerequisite to any determination of planting stock quality. Not only does performance have to be determined, but performance has to be rated against the objectives of management. Planting stock is produced in order to give effect to the forest policy of the organization.

A distinction needs to be made between "planting stock quality" and "planting stock performance potential" (PSPP). The actual performance of any given batch of outplanted planting stock is determined only in part by the kind and condition, i.e., the intrinsic PSPP, of the planting stock.

The PSPP is impossible to estimate reliably by eye because outward appearance, especially of stock withdrawn from refrigerated storage, can deceive even experienced foresters, who would be offended if their ability were questioned to recognize good planting stock when they saw it. Prior to Wakeley's (1954) demonstration of the importance of the physiological state of planting stock in determining the ability of the stock to perform after outplanting, and to a considerable extent even afterwards, morphological appearance has generally served as the basis for estimating the quality of planting stock. Gradually, however, a realization developed that more was involved. Tucker et al. (1968), for instance, after assessing 10-year survival data from several experimental white spruce plantations in Manitoba noted that "Perhaps the most important point revealed here is that certain lots of transplants performed better than others", even though all transplants were handled and planted with care. The intuitive "stock that looks good must be good" is a persuasive, but potentially dangerous maxim. That greatest of teachers, Bitter Experience, has often enough demonstrated the fallibility of such assessment, even though the corollary "stock that looks bad must be bad" is likely to be well founded. The physiological qualities of planting stock are hidden from the eye and must be revealed by testing. The potential for survival and growth of a batch of planting stock may be estimated from various features, morphological and physiological, of the stock or a sample thereof.

The size and shape and general appearance of a seedling can nevertheless give useful indications of PSPP. In low-stress outplanting situations, and with a minimized handling and lifting-planting cycle, a system based on specification for nursery stock and minimum morphological standards for acceptable seedlings works tolerably well. In certain circumstances, benefits often accrue from the use of large planting stock of highly ranked morphological grades. Length of leading shoot, diameter of stem, volume of root system, shoot:root ratios, and height:diameter ratios have been correlated with performance under specific site and planting conditions. However, the concept that larger is better negates the underlying complexities. Schmidt-Vogt (1980), for instance, found that whereas mortality among large outplants is greater than among small in the year of planting, mortality in subsequent growing seasons is higher among small outplants than among large. Much of the literature on comparative seedling performance is clouded by uncertainty as to whether the stocks being compared share the same physiological condition; differences invalidate such comparisons.

Height and root-collar diameter are generally accepted as the most useful morphological criteria and are often the only ones used in specifying standards.Quantification of root system morphology is difficult but can be done, e.g. by using the photometric rhizometer to determine intercept area, or volume by displacement or gravimetric methods.

Planting stock is always subject to a variety of conditions that are never optimal *in toto*. The effect of sub-optimal conditions is to induce stress in the plants. The nursery manager aims, and is normally able to avoid stresses greater than moderate, i.e., restricting stresses to levels that can be tolerated by the plants without incurring serious damage. The adoption of nursery regimes to equip planting stock with characteristics conferring increased ability to withstand outplanting stresses, by managing stress levels in the nursery to "condition" planting stock to increase tolerance to various post-planting environmental stresses, has become widespread, particularly with containerized stock.

Outplanted stock that is unable to tolerate high temperatures occurring at soil surfaces will fail to establish on many forest sites, even in the far north. Factors affecting heat tolerance were investigated by Colombo et al. (1995); the production and roles of heat shock proteins (HSPs) are important in this regard. HSPs, present constitutively in black spruce and many other, perhaps most, higher plants are important both for normal cell functioning and in a stress response mechanism following exposure to high, non-lethal temperature. In black spruce at least, there is an association between HSPs and increased levels of heat tolerance. Investigation of the diurnal variability in heat tolerance of roots and shoots in black spruce seedlings 14 to 16 weeks old found in all 4 trials that shoot heat tolerance was significantly greater in the afternoon than in the morning. The trend in root heat tolerance was similar to that found in the shoots; root systems exposed to 47 °C for 15 minutes in the afternoon averaged 75 new roots after a 2-week growth period, whereas only 28 new roots developed in root systems similarly exposed in the morning. HSP_{73} was detected in black spruce nuclear, mitochondrial, microsomal, and soluble protein fractions, while HSP_{72} was observed only in the soluble protein fraction. Seedlings exhibited constitutive synthesis of HSP_{73} at 26 °C in all except the nuclear membrane fraction in the morning; HSP levels at 26 °C in the afternoon were higher than in the morning in the mitochondrial and microsomal protein factions. Heat shock affected the abundance of HSPs depending on protein fraction and time of day. Without heat shock, nuclear membrane-bound HSP_{73} was absent from plants in the morning and only weakly present in the afternoon, and heat shock increased the abundance of nuclear membrane. Heat shock also affected the abundance of HSP_{73} in the afternoon, and caused HSP_{73} to appear in the morning. In the mitochondrial and microsomal protein fractions, an afternoon heat shock reduced HSP_{73}, whereas a morning heat shock increased HSP_{73} in the mitochondrial but decreased it in the microsomal fraction. Heat shock increased soluble $HSP_{72/73}$ levels in both the morning and afternoon. In all instances, shoot and root heat tolerances were significantly greater in the afternoon than in the morning.

Planting stock continues to respire during storage even if frozen. Temperature is the major factor controlling the rate, and care must be taken to avoid overheating. Navratil (1982) found that closed containers in cold storage averaged internal temperatures 1.5 °C to 2.0 °C above the nominal storage temperature. Depletion of reserves can be estimated from the decrease in dry weight. Cold-stored 3+0 white spruce nursery stock in northern Ontario had lost 9% to 16% of dry weight after 40 days of storage. Carbohydrates can also be determined directly.

The propensity of a root system to develop new roots or extend existing roots cannot be determined by eye, yet it is the factor that makes or breaks the outcome of an outplanting operation. The post-planting development of roots or root systems of coniferous planting stock is determined by many factors, some physiological, some environmental. Unsatisfactory rates of post-planting survival unrelated to the morphology of the stock, led to attempts to test the physiological condition of planting stock, particularly to quantify the propensity to produce new root growth. New root growth can be assumed to be necessary for successful establishment of stock after planting, but although the thesis that RGC is positively related to field performance would seem to be reasonable, supporting evidence has been meager.

The physiological condition of seedlings is reflected by changes in root activity. This is helpful in determining the readiness of stock for lifting and storing and also for outplanting after storage. Navratil (1982) reported a virtually perfect ($R^2 = 0.99$) linear relationship in the frequency of 3+0 white spruce white root tips longer than 10 mm with time in the fall at Pine Ridge Forest Nursery, Alberta, decreasing during a 3-week period to zero on October 13 in 1982.Root regenerating research with white spruce in Canada (Hambly 1973, Day and MacGillivray 1975, Day and Breunig 1997) followed similar lines to that of Stone's (1955) pioneering work in California.

Simpson and Ritchie (1997) debated the proposition that root growth potential of planting stock predicts field performance; their conclusion was that root growth potential, as a surrogate for seedling vigor, can predict field performance, but only under such situations as site conditions permit. Survival after planting is only partly a function of an outplant's ability to initiate roots in test conditions; root growth capacity is not the sole predictor of plantation performance.

Some major problems militate against greater use of RGC in forestry, including: unstandardized techniques; unstandardized quantification; uncertain correlation between quantified RGC and field performance; variability within given, nominally identical, kinds of planting stock; and the irrelevance of RGC test values determined on a sub-sample of a parent population that subsequently, before it is planted, undergoes any substantive physiological or physical change. In its present form, RGC testing is silviculturally useful chiefly as a means of detecting planting stock that, while visually unimpaired, is moribund.

Seedling moisture content can be increased or decreased in storage, depending on various factors including especially the type of container and the kind and amount of moisture-retaining material present. When seedlings exceed 20 bars PMS in storage, survival after outplanting becomes problematical. The Relative Moisture Content of stock lifted during dry conditions can be increased gradually when stored in appropriate conditions. White spruce (3+0) packed in Kraft bags in northern Ontario increased RMC by 20% to 36% within 40 days.

Bareroot 1.5+1.5 white spruce were taken from cold storage and planted early in May on a clear-felled boreal forest site in northeastern Ontario. Similar plants were potted and kept in a greenhouse. In outplanted trees, maximum stomatal conductances (g) were initially low (<0.01 cm/s), and initial base xylem pressure potentials (PSIb) were -2.0 MPa. During the growing season, g increased to about 0.20 cm/s and PSIb to -1.0 MPa. Minimum xylem pressure potential (PSIm) was initially -2.5 MPa, increasing to -2.0 MPa on day 40, and about -1.6 MPa by day 110. During the first half of the growing season, PSIm was below turgor loss point. The osmotic potential at turgor loss point decreased after planting to -2.3 MPa 28 days later. In the greenhouse, minimum values of PSIT were -2.5 MPa (in the first day after planting. the maximum bulk modulus of elasticity was greater in white spruce than in similarly treated jack pine and showed greater seasonal changes. Relative water content (RWC) at turgor loss was 80-87%. Available turgor (TA), defined as the integral of turgor over the range of RWC between PSIb and xylem pressure potential at the turgor loss point) was 4.0% for white spruce at the beginning of the season compared with 7.9% for jack pine, but for the rest of the season TA for jack pine was only 2%, to 3% that of white spruce. Diurnal turgor (Td), the integral of turgor over the range of RWC between PSIb and PSIm, as a percentage of TA was higher in field-planted white spruce than jack pine until the end of the season.

The stomata of both white and black spruce were more sensitive to atmospheric evaporative demands and plant moisture stress during the first growing season after outplanting on 2 boreal sites in northern Ontario than were jack pine stomata, physiological differences that favoured growth and establishment being more in jack pine than in the spruces.

With black spruce and jack pine, but not with white spruce, Grossnickle and Blake's (1987) findings warrant mention in relation to the bareroot-containerized debate. During the first growing season after outplanting, containerized seedlings of both species had greater needle conductance than bareroot seedlings over a range of absolute humidity deficits. Needle conductance of containerized seedlings of both species remained high during periods of high absolute humidity deficits and increasing plant moisture stress. Bareroot outplants of both species had a greater early season resistance to water-flow through the soil–plant–atmosphere continuum (SPAC) than had containerized outplants. Resistance to water flow through the SPAC decreased in bareroot stock of both species as the season progressed, and was comparable to containerized seedlings 9 to 14 weeks after planting. Bareroot black spruce had greater new-root development than containerized stock throughout the growing season.

The greater efficiency of water use in newly transplanted 3-year-old white spruce seedlings under low levels of absolute humidity difference in water-stressed plants immediately after planting helps explain the commonly observed favourable response of young outplants to the nursing effect of a partial canopy. Silvicultural treatments promoting higher humidity levels at the planting microsite should improve white spruce seedling photosynthesis immediately after planting.

Stock Types (Seedling Nomenclature)

Planting stock is grown under many diverse nursery culture regimes, in facilities ranging from sophisticated computerized greenhouses to open compounds. Types of stock include bareroot seedlings and transplants, and various kinds of containerized stock. For simplicity, both container-grown and bareroot stock are generally referred to as seedlings, and transplants are nursery stock that have been lifted and transplanted into another nursery bed, usually at wider spacing. The size and physiological character of stock vary with the length of growing period and with growing conditions. Until the technology of raising containerized nursery stock bourgeoned in the second half of the twentieth-century, bareroot planting stock classified by its age in years was the norm.

Classification by Age

The number of years spent in the nursery seedbed by any particular lot of planting stock is indicated by the 1st of a series of numbers. The 2nd number indicates the years subsequently spent in the transplant line, and a zero is shown if indeed there has been no transplanting. A 3rd number, if any, would indicate the years subsequently spent after a second lifting and transplanting. The numbers are sometimes separated by dashes, but separation by plus sign is more logical inasmuch as the sum of the individual numbers gives the age of the planting stock. Thus 2+0 is 2-year-old seedling planting stock that has not been transplanted, and Candy's (1929) white spruce 2+2+3 stock had spent 2 years in the seedbed, 2 years in transplant lines, and another 3 years in transplant lines after a second transplanting. Variations have included such self-explanatory combinations, such as 1½+1½, etc.

The class of planting stock to use on a particular site is generally selected on the basis of historical record of survival, growth, and total cost of surviving trees. In the Lake States, Kittredge concluded that good stock of 2+1 white spruce was the smallest size likely to succeed and was better than larger and more expensive stock when judged by final cost of surviving trees.

Classification by Seedling Description Code

Because age alone is an inadequate descriptor of planting stock, various codes have been developed to describe such components of stock characteristics as height, stem diameter, and shoot:-root ratio. A description code may include an indication of the intended planting season.

Physiological Characteristics

Neither age classification nor seedling description code indicate the physiological condition of planting stock, though rigid adherence to a given cultural regime together with observation of performance over a number of years of planting can produce stock suitable for performing on a "same again" basis.

Classification by System

Planting stock is raised under a variety of systems, but these have devolved generally into 2 main groupings: bareroot and containerized. Manuals specifically for the production of bareroot and containerized nursery stock are valuable resources for the nursery manager. As well, a lot of good information about nursery stock specific to regional jurisdictions is well presented by Cleary et al. (1978) for Oregon, Lavender et al. (1990) for British Columbia, and Wagner and Colombo (2001) for Ontario.

Plantation

A pine plantation in the United States. Tree plantations are usually easily distinguished from natural forests by the trees being planted in straight lines.

One operator planting poplar cuttings with a Rotor machine for planting a new nursery.

A plantation is a large piece of land (or water) where one crop is specifically planted for widespread commercial sale and usually tended by resident laborers. The crops grown include fast-growing trees (often conifers), cotton, coffee, tea, cocoa, sugar cane, sisal, oil seeds (e.g. oil palms), rubber trees, and various fruits. Protectionist policies and natural comparative advantage have sometimes contributed to determining where plantations were located.

Among the earliest examples of plantations were the latifundia of the Roman Empire, which produced large quantities of wine and olive oil for export. Plantation agriculture grew rapidly with the increase in international trade and the development of a worldwide economy that followed the expansion of European colonial empires. Like every economic activity, it has changed over time. Earlier forms of plantation agriculture were associated with large disparities of wealth and income, foreign ownership and political influence, and exploitative social systems such as indentured labour and slavery.

Forestry

Industrial

Industrial plantations are established to produce a high volume of wood in a short period of time. Plantations are grown by state forestry authorities (for example, the Forestry Commission in Britain) and/or the paper and wood industries and other private landowners (such as Weyerhaeuser, Rayonier and Plum Creek Timber in the United States, Asia Pulp & Paper in Indonesia). Christmas trees are often grown on plantations as well. In southern and southeastern Asia, teak plantations have recently replaced the natural forest.

A plantation of Douglas-fir in Washington, U.S.

Machine for industrial plantations

Industrial plantations are actively managed for the commercial production of forest products. Industrial plantations are usually large-scale. Individual blocks are usually even-aged and often consist of just one or two species. These species can be exotic or indigenous. The plants used for the plantation are often genetically altered for desired traits such as growth and resistance to pests and diseases in general and specific traits, for example in the case of timber species, volumic wood production and stem straightness. Forest genetic resources are the basis for genetic alteration. Selected individuals grown in seed orchards are a good source for seeds to develop adequate planting material.

A stump grinder working to remove a tree stump.

Wood production on a tree plantation is generally higher than that of natural forests. While forests managed for wood production commonly yield between 1 and 3 cubic meters per hectare per year, plantations of fast-growing species commonly yield between 20 and 30 cubic meters or more per hectare annually; a Grand Fir plantation at Craigvinean in Scotland has a growth rate of 34 cubic meters per hectare per year (Aldhous & Low 1974), and Monterey Pine plantations in southern Australia can yield up to 40 cubic meters per hectare per year (Everard & Fourt 1974). In 2000, while plantations accounted for 5% of global forest, it is estimated that they supplied about 35% of the world's roundwood.

Growth Cycle

- In the first year, the ground is prepared usually by the combination of burning, herbicide spraying, and/or cultivation and then saplings are planted by human crew or by machine. The saplings are usually obtained in bulk from industrial nurseries, which may specialize in selective breeding in order to produce fast growing disease- and pest-resistant strains.
- In the first few years until the canopy closes, the saplings are looked after, and may be dusted or sprayed with fertilizers or pesticides until established.
- After the canopy closes, with the tree crowns touching each other, the plantation is becoming dense and crowded, and tree growth is slowing due to competition. This stage is termed 'pole stage'. When competition becomes too intense (for pine trees, when the live crown is less than a third of the tree's total height), it is time to thin out the section. There are several methods for thinning, but where topography permits, the most popular is 'row-thinning', where every third or fourth or fifth row of trees is removed, usually with a harvester. Many trees are removed, leaving regular clear lanes through the section so that the remaining trees have room to expand again. The removed trees are delimbed, forwarded to the forest road, loaded onto trucks, and sent to a mill. A typical pole stage plantation tree is 7–30 cm in diameter at breast height (dbh). Such trees are sometimes not suitable for timber, but are used as pulp for paper and particleboard, and as chips for oriented strand board.
- As the trees grow and become dense and crowded again, the thinning process is repeated.

Depending on growth rate and species, trees at this age may be large enough for timber milling; if not, they are again used as pulp and chips.

- Around year 10-60 the plantation is now mature and (in economic terms) is falling off the back side of its growth curve. That is to say, it is passing the point of maximum wood growth per hectare per year, and so is ready for the final harvest. All remaining trees are felled, delimbed, and taken to be processed.
- The ground is cleared, and the cycle is repeated.

Some plantation trees, such as pines and eucalyptus, can be at high risk of fire damage because their leaf oils and resins are flammable to the point of a tree being explosive under some conditions. Conversely, an afflicted plantation can in some cases be cleared of pest species cheaply through the use of a prescribed burn, which kills all lesser plants but does not significantly harm the mature trees.

Natural Forest Loss

Many forestry experts claim that the establishment of plantations will reduce or eliminate the need to exploit natural forest for wood production. In principle this is true because due to the high productivity of plantations less land is needed. Many point to the example of New Zealand, where 19% of the forest area provides 99% of the supply of industrial round wood. It has been estimated that the world's demand for fiber could be met by just 5% of the world forest (Sedjo & Botkin 1997). However, in practice, plantations are replacing natural forest, for example in Indonesia. According to the FAO, about 7% of the natural closed forest being lost in the tropics is land being converted to plantations. The remaining 93% of the loss is land being converted to agriculture and other uses. Worldwide, an estimated 15% of plantations in tropical countries are established on closed canopy natural forest.

In the Kyoto Protocol, there are proposals encouraging the use of plantations to reduce carbon dioxide levels (though this idea is being challenged by some groups on the grounds that the sequestered CO_2 is eventually released after harvest).

A tea plantation in Ciwidey, Bandung in Indonesia

Criticism

In contrast to a naturally regenerated forest, plantations are typically grown as even-aged monocultures, primarily for timber production.

- Plantations are usually near- or total monocultures. That is, the same species of tree is planted across a given area, whereas a natural forest would contain a far more diverse range of tree species.

- Plantations may include tree species that would not naturally occur in the area. They may include unconventional types such as hybrids, and genetically modified trees may be used sometime in the future. Since the primary interest in plantations is to produce wood or pulp, the types of trees found in plantations are those that are best-suited to industrial applications. For example, pine, spruce and eucalyptus are widely planted far beyond their natural range because of their fast growth rate, tolerance of rich or degraded agricultural land and potential to produce large volumes of raw material for industrial use.

- Plantations are always young forests in ecological terms. Typically, trees grown in plantations are harvested after 10 to 60 years, rarely up to 120 years. This means that the forests produced by plantations do not contain the type of growth, soil or wildlife typical of old-growth natural forest ecosystems. Most conspicuous is the absence of decaying dead wood, a crucial component of natural forest ecosystems.

Bushfires pose a high risk to Eucalyptus plantations.

In the 1970s, Brazil began to establish high-yield, intensively managed, short rotation plantations. These types of plantations are sometimes called fast-wood plantations or fiber farms and often managed on a short-rotation basis, as little as 5 to 15 years. They are becoming more widespread in South America, Asia and other areas. The environmental and social impacts of this type of plantation has caused them to become controversial. In Indonesia, for example, large multi-national pulp companies have harvested large areas of natural forest without regard for regeneration. From 1980 to 2000, about 50% of the 1.4 million hectares of pulpwood plantations in Indonesia have been established on what was formerly natural forest land.

The replacement of natural forest with tree plantations has also caused social problems. In some countries, again, notably Indonesia, conversions of natural forest are made with little regard for rights of the local people. Plantations established purely for the production of fiber provide a much narrower range of services than the original natural forest for the local people. India has sought to limit this damage by limiting the amount of land owned by one entity and, as a result, smaller plantations are owned by local farmers who then sell the wood to larger companies. Some large environmental organizations are critical of these high-yield plantations and are running an anti-plantation campaign, notably the Rainforest Action Network and Greenpeace.

Farm and Home

Farm or home plantations are typically established for the production of timber and fire wood for home use and sometimes for sale. Management may be less intensive than with Industrial plantations. In time, this type of plantation can become difficult to distinguish from naturally regenerated forest.

Teak and bamboo plantations in India have given good results and an alternative crop solution to farmers of central India, where conventional farming was popular. But due to rising input costs of farming many farmers have done teak and bamboo plantations which require very little water (only during first two years). Teak and bamboo have legal protection from theft. Bamboo, once planted, gives output for 50 years till flowering occurs. Teak requires 20 years to grow to full maturity and fetch returns.

Environmental

These may be established for watershed or soil protection. They are established for erosion control, landslide stabilization and windbreaks. Such plantations are established to foster native species and promote forest regeneration on degraded lands as a tool of environmental restoration.

Ecological Impact

The last batch of sawnwood from the peat forest in Indragiri Hulu, Sumatra, Indonesia. Deforestation for oil palm plantation.

Probably the single most important factor a plantation has on the local environment is the site where the plantation is established. If natural forest is cleared for a planted forest then a reduction in biodiversity and loss of habitat will likely result. In some cases, their establishment may involve draining wetlands to replace mixed hardwoods that formerly predominated with pine species. If a plantation is established on abandoned agricultural land, or highly degraded land, it can result in an increase in both habitat and biodiversity. A planted forest can be profitably established on lands that will not support agriculture or suffer from lack of natural regeneration.

The tree species used in a plantation is also an important factor. Where non-native varieties or species are grown, few of the native fauna are adapted to exploit these and further biodiversity loss occurs. However, even non-native tree species may serve as corridors for wildlife and act as a buffer for native forest, reducing edge effect.

Once a plantation is established, how it is managed becomes the important environmental factor. The single most important factor of management is the rotation period. Plantations harvested on longer rotation periods (30 years or more) can provide similar benefits to a naturally regenerated forest managed for wood production, on a similar rotation. This is especially true if native species are used. In the case of exotic species, the habitat can be improved significantly if the impact is mitigated by measures such as leaving blocks of native species in the plantation, or retaining corridors of natural forest. In Brazil, similar measures are required by government regulation

Sugar

Sugar plantations were highly valued in the Caribbean by the British and French colonists in the 17th and 18th centuries and the use of sugar in Europe rose during this period. Sugarcane is still an important crop in Cuba. Sugar plantations also arose in countries such as Barbados and Cuba because of the natural endowments that they had. These natural endowments included soil that was conducive to growing sugar and a high marginal product of labour realized through the increasing number of slaves.

Rubber

Sugarcane plantation in rural Cuba

Plantings of para rubber, the tree *Hevea brasiliensis*, are usually called plantations.

Oil Palm

Oil palm agriculture is rapidly expanding across wet tropical regions, and is usually developed at plantation scale.

Orchards

Fruit orchards are sometimes considered to be plantations.

Arable Crops

These include tobacco, sugarcane, pineapple, and cotton, especially in historical usage.

Before the rise of cotton in the American South, indigo and rice were also sometimes called plantation crops.

Harvesting tea in Bogor, West Java

Fishing

When Newfoundland was colonized by England in 1610, the original colonists were called "Planters" and their fishing rooms were known as "fishing plantations". These terms were used well into the 20th century.

The following three plantations are maintained by the Government of Newfoundland and Labrador as provincial heritage sites:

- Sea-Forest Plantation was a 17th-century fishing plantation established at Cuper's Cove (present-day Cupids) under a royal charter issued by King James I.
- Mockbeggar Plantation is an 18th-century fishing plantation at Bonavista.
- Pool Plantation a 17th-century fishing plantation maintained by Sir David Kirke and his heirs at Ferryland. The plantation was destroyed by French invaders in 1696.

Other fishing plantations:

- Bristol's Hope Plantation, a 17th-century fishing plantation established at Harbour Grace, created by the Bristol Society of Merchant-Adventurers.
- Benger Plantation, an 18th-century fishing plantation maintained by James Benger and his heirs at Ferryland. It was built on the site of Georgia plantation.

- Piggeon's Plantation, an 18th-century fishing plantation maintained by Ellias Piggeon at Ferryland.

Slavery and Para-slavery

1913 photo: blacks picking cotton on a plantation in the South

Slave labor extracted from imported Africans was used extensively to work on early plantations (such as tobacco, rice, cotton, and sugar plantations) in the American colonies and the United States, throughout the Caribbean, the Americas, and in European-occupied areas of Africa. Several notable historians and economists such as Eric Williams, Walter Rodney, and Karl Marx contend that the global capitalist economy was largely founded upon the creation and produce of thousands of slave labour camps based in colonial plantations, exploiting tens of millions of purchased Africans.

Italian immigrants working on Brazilian coffee plantation, early 20th century

In modern times, the low wages typically paid to plantation workers are the basis of plantation profitability in some areas. Sugarcane plantations in the Caribbean and Brazil, worked by slave labour, were also examples of the plantation system.

In more recent times, overt slavery has been replaced by *para-slavery* or *slavery-in-kind*, including the sharecropping system. At its most extreme, workers are in "debt bondage": they must work to pay off a debt at such punitive interest rates that it may never be paid off. Others work unreasonably long hours and are paid subsistence wages that (in practice) may only be spent in the company store.

In Brazil, a sugarcane plantation was termed an *engenho* ("engine"), and the 17th-century English usage for organized colonial production was "factory." Such colonial social and economic structures are discussed at Plantation economy.

Sugar workers on plantations in Cuba and elsewhere in the Caribbean lived in company towns known as *bateyes*.

Antebellum American South

In the American South, antebellum plantations were centered on a "plantation house," the residence of the owner, where important business was conducted. Slavery and plantations had different characteristics in different regions of the South. As the Upper South of the Chesapeake Bay colonies developed first, historians of the antebellum South defined planters as those who held 20 or more slaves. Major planters held many more, especially in the Deep South as it developed. The majority of slaveholders held 10 or fewer slaves, often just a few to labor domestically. By the late 18th century, most planters in the Upper South had switched from exclusive tobacco cultivation to mixed-crop production, both because tobacco had exhausted the soil and because of changing markets. The shift away from tobacco meant they had slaves in excess of the number needed for labor, and they began to sell them in the internal slave trade.

Old Jackson Plantation house, Schriever, Louisiana, June 1940

There was a variety of domestic architecture on plantations. The largest and wealthiest planter families, for instance, those with estates fronting on the James River in Virginia, constructed mansions in brick and Georgian style, e.g. Shirley Plantation. Common or smaller planters in the late 18th and 19th century had more modest wood frame buildings, such as Southall Plantation in Charles City County.

In the Lowcountry of South Carolina, by contrast, even before the American Revolution, planters holding large rice plantations typically owned hundreds of slaves. In Charleston and Savannah, the elite also held numerous slaves to work as household servants. The 19th-century development of the Deep South for cotton cultivation depended on large plantations with much more acreage than was typical of the Upper South; and for labor, planters held hundreds of slaves.

Until December 1865 slavery was legal in parts of the United States. Most slaves were employed in agriculture, and *planter* was a term commonly used to describe a farmer with many slaves.

The term planter has no universally-accepted definition, but academic historians have defined it to identify the elite class, "a landowning farmer of substantial means." In the "Black Belt" counties of Alabama and Mississippi, the terms "planter" and "farmer" were often synonymous. Robert Fogel and Stanley Engerman define large planters as owning over 50 slaves, and medium planters as owning between 16 and 50 slaves.

In his study of Black Belt counties in Alabama, Jonathan Wiener defines planters by ownership of real property, rather than of slaves. A planter, for Wiener, owned at least $10,000 worth of real estate in 1850 and $32,000 worth in 1860, equivalent to about the top 8 percent of landowners. In his study of southwest Georgia, Lee Formwalt also defines planters in size of land holdings rather than slaves. Formwalt's planters are in the top 4.5 percent of land owners, translating into real estate worth $6,000 or more in 1850, $24,000 or more in 1860, and $11,000 or more in 1870. In his study of Harrison County, Texas, Randolph B. Campbell classifies large planters as owners of 20 slaves, and small planters as owners of between ten and 19 slaves. In Chicot and Phillips counties, Arkansas, Carl H. Moneyhon defines large planters as owners of twenty or more slaves, and six hundred or more acres.

Forest Gardening

Forest gardening is a low maintenance sustainable plant-based food production and agroforestry system based on woodland ecosystems, incorporating fruit and nut trees, shrubs, herbs, vines and perennial vegetables which have yields directly useful to humans. Making use of companion planting, these can be intermixed to grow in a succession of layers, to build a woodland habitat.

Robert Hart's forest garden in Shropshire

Forest gardening is a prehistoric method of securing food in tropical areas. In the 1980s, Robert Hart coined the term "forest gardening" after adapting the principles and applying them to te perate climates.

History

Forest gardens are probably the world's oldest form of land use and most resilient agroecosystem. They originated in prehistoric times along jungle-clad river banks and in the wet foothills of monsoon regions. In the gradual process of families improving their immediate environment, useful tree and vine species were identified, protected and improved whilst undesirable species were eliminated. Eventually superior foreign species were selected and incorporated into the gardens.

Forest gardens are still common in the tropics and known by various names such as: *home gardens* in Kerala in South India, Nepal, Zambia, Zimbabwe and Tanzania; *Kandyan forest gardens* in Sri Lanka; *huertos familiares*, the "family orchards" of Mexico; and *pekarangan*, the gardens of "complete design", in Java. These are also called agroforests and, where the wood components are short-statured, the term shrub garden is employed. Forest gardens have been shown to be a significant source of income and food security for local populations.

Robert Hart adapted forest gardening for the United Kingdom's temperate climate during the 1980s. His theories were later developed by Martin Crawford from the Agroforestry Research Trust and various permaculturalists such as Graham Bell, Patrick Whitefield, Dave Jacke and Geoff Lawton.

In Tropical Climates

Forest gardens, or home gardens, are common in the tropics, using intercropping to cultivate trees, crops, and livestock on the same land. In Kerala in south India as well as in northeastern India, the home garden is the most common form of land use and is also found in Indonesia. One example combines coconut, black pepper, cocoa and pineapple. These gardens exemplify polyculture, and conserve much crop genetic diversity and heirloom plants that are not found in monocultures. Forest gardens have been loosely compared to the religious concept of the Garden of Eden.

Americas

The BBC's *Unnatural Histories* claimed that the Amazon rainforest, rather than being a pristine wilderness, has been shaped by humans for at least 11,000 years through practices such as forest gardening and *terra preta*. This was also explored in the bestselling book *1491* by author Charles C. Mann. Since the 1970s, numerous geoglyphs have also been discovered on deforested land in the Amazon rainforest, furthering the evidence about Pre-Columbian civilizations.

On the Yucatán Peninsula, much of the Maya food supply was grown in "orchard-gardens", known as *pet kot*. The system takes its name from the low wall of stones (*pet* meaning circular and *kot* wall of loose stones) that characteristically surrounds the gardens.

Africa

In many African countries, for example Zambia, Zimbabwe, Ethiopia and Tanzania, gardens are widespread in rural, periurban and urban areas and they play an essential role in establishing food security. Most well known are the Chaga or Chagga gardens on the slopes of Mt. Kilimanjaro in Tanzania. These are an excellent example of an agroforestry system. In many countries, women are the main actors in home gardening and food is mainly produced for subsistence. In North-Africa, oasis layered gardening with palm trees, fruit trees and vegetables is a traditional type of forest garden.

Nepal

In Nepal, the *Ghar Bagaincha*, literally "home garden", refers to the traditional land use system around a homestead, where several species of plants are grown and maintained by household members and their products are primarily intended for the family consumption (Shrestha et al., 2002). The term "home garden" is often considered synonymous to the kitchen garden. However, they differ in terms of function, size, diversity, composition and features (Sthapit et al., 2006). In Nepal, 72% of households have home gardens of an area 2–11% of the total land holdings (Gautam et al., 2004). Because of their small size, the government has never identified home gardens as an important unit of food production and they thereby remain neglected from research and development. However, at the household level the system is very important as it is an important source of quality food and nutrition for the rural poor and, therefore, are important contributors to the household food security and livelihoods of farming communities in Nepal. The gardens are typically cultivated with a mixture of annual and perennial plants that can be harvested on a daily or seasonal basis. Biodiversity that has an immediate value is maintained in home gardens as women and children have easy access to preferred food. Home gardens, with their intensive and multiple uses, provide a safety net for households when food is scarce. These gardens are not only important sources of food, fodder, fuel, medicines, spices, herbs, flowers, construction materials and income in many countries, they are also important for the in situ conservation of a wide range of unique genetic resources for food and agriculture (Subedi et al., 2004). Many uncultivated, as well as neglected and underutilised species could make an important contribution to the dietary diversity of local communities (Gautam et al., 2004).

In addition to supplementing diet in times of difficulty, home gardens promote whole-family and whole-community involvement in the process of providing food. Children, the elderly, and those caring for them can participate in this infield agriculture, incorporating it with other household tasks and scheduling. This tradition has existed in many cultures around the world for thousands of years.

In Mediterranean Climates

The mediterranean climate has long, hot, rainless summers and relatively short, cool, rainy winters (Köppen climate classification *Csa*). Its climate conditions are highly variable within an area and modified locally by altitude, latitude, and the proximity to the Mediterranean. In the 1950s the Forest Research Department of the Ministry of Agriculture founded a botanical forest garden in the Sharon region in Israel, the Ilanot Forest. As the only one of its kind in Israel, it harbours more than seven hundred and fifty species of trees from locations all over the world, including the Japanese sago palm cycas revoluta, fig trees (ficus glomerata), stone pine trees (pinus pinea) that produce tasty pine nuts and adds to the biodiversity of Israel.

In Temperate Climates

Robert Hart, forest gardening pioneer

Robert Hart coined the term "forest gardening" during the 1980s. Hart began farming at Wenlock Edge in Shropshire with the intention of providing a healthy and therapeutic environment for himself and his brother Lacon. Starting as relatively conventional smallholders, Hart soon discovered that maintaining large annual vegetable beds, rearing livestock and taking care of an orchard were tasks beyond their strength. However, a small bed of perennial vegetables and herbs he planted was looking after itself with little intervention.

Following Hart's adoption of a raw vegan diet for health and personal reasons, he replaced his farm animals with plants. The three main products from a forest garden are fruit, nuts and green leafy vegetables. He created a model forest garden from a 0.12 acre (500 m^2) orchard on his farm and intended naming his gardening method *ecological horticulture* or *ecocultivation*. Hart later dropped these terms once he became aware that *agroforestry* and *forest gardens* were already being used to describe similar systems in other parts of the world. He was inspired by the forest farming methods of Toyohiko Kagawa and James Sholto Douglas, and the productivity of the Keralan home gardens as Hart explains:

From the agroforestry point of view, perhaps the world's most advanced country is the Indian state of Kerala, which boasts no fewer than three and a half million forest gardens...As an example of the extraordinary intensivity of cultivation of some forest gardens, one plot of only 0.12 hectares (0.30 acres) was found by a study group to have twenty-three young coconut palms, twelve cloves, fifty-six bananas, and forty-nine pineapples, with thirty pepper vines trained up its trees. In addition, the small holder grew fodder for his house-cow.

Seven-layer System

Robert Hart pioneered a system based on the observation that the natural forest can be divided into distinct levels. He used intercropping to develop an existing small orchard of apples and pears into an edible polyculture landscape consisting of the following layers:

1. 'Canopy layer' consisting of the original mature fruit trees.
2. 'Low-tree layer' of smaller nut and fruit trees on dwarfing root stocks.

3. 'Shrub layer' of fruit bushes such as currants and berries.
4. 'Herbaceous layer' of perennial vegetables and herbs.
5. 'Rhizosphere' or 'underground' dimension of plants grown for their roots and tubers.
6. 'Ground cover layer' of edible plants that spread horizontally.
7. 'Vertical layer' of vines and climbers.

A key component of the seven-layer system was the plants he selected. Most of the traditional vegetable crops grown today, such as carrots, are sun loving plants not well selected for the more shady forest garden system. Hart favoured shade tolerant perennial vegetables.

The seven layers of the forest garden

Further Development

The Agroforestry Research Trust (ART), managed by Martin Crawford, runs experimental forest gardening projects on a number of plots in Devon, United Kingdom. Crawford describes a forest garden as a low-maintenance way of sustainably producing food and other household products.

Ken Fern had the idea that for a successful temperate forest garden a wider range of edible shade tolerant plants would need to be used. To this end, Fern created the organisation Plants for a Future (PFAF) which compiled a plant database suitable for such a system. Fern used the term *woodland gardening*, rather than forest gardening, in his book *Plants for a Future.*

The Movement for Compassionate Living (MCL) promote forest gardening and other types of vegan organic gardening to meet society's needs for food and natural resources. Kathleen Jannaway, the founder of MCL, wrote a book outlining a sustainable vegan future called *Abundant Living in the Coming Age of the Tree* in 1991. In 2009, the MCL provided a grant of £1,000 to the Bangor Forest Garden project in Gwynedd, North West Wales.

Kevin Bradley coined the phrase "Edible Forest" in the 1980s as the name of his nursery, garden, and orchard on 5 acres in the frigid zone 3 pine forests of northern Wisconsin. Among 3 options, he chose "Edible Forest" because it "evokes at once an ethereal, spiritual, and magical image", of Disney- like "Forest of No Return"; of the biblical "Garden of Eden". This image was perfectly in line with his ongoing experiment begun in 1985 in what he calls a closed loop human environment, combining multi- story tree and field crop "garden/orchards" for maximum beauty and use of space, someday to be very useful in an ever shrinking world. "The name, at the same time, with its irrational first impression (of course we can't eat a forest), forces the mind to think, if just a little bit, about its inference and thus sticks in our memories". It appeared from Bradley's research that the two words had, prior to the 80's, never been put together before as a noun phrase but which by today, after more than two decades of Bradley's "Edible Forest Nursery" and the 2005 text by Jacke and Toensmeirer's- "Edible Forest Gardens", has grown into a movement and little "Edible Forests" all over the world.

In 2005, Dave Jacke and Eric Toensmeier's two-volume *Edible Forest Gardens* provided a deeply researched reference focused on North American forest gardening climates, habitats, and species. The book attempts to ground forest gardening deeply in ecological science. The Apios Institute wiki grew out of their work, and seeks to document and share the experience of people around the world working with the species in polycultures.

Permaculture

Bill Mollison, who coined the term *permaculture*, visited Robert Hart at his forest garden in Wenlock Edge in October 1990. Hart's seven-layer system has since been adopted as a common permaculture design element.

Numerous permaculturalists are proponents of forest gardens, or food forests, such as Graham Bell, Patrick Whitefield, Dave Jacke, Eric Toensmeier and Geoff Lawton. Bell started building his forest garden in 1991 and wrote the book *The Permaculture Garden* in 1995, Whitefield wrote the book *How to Make a Forest Garden* in 2002, Jacke and Toensmeier co-authored the two volume book set *Edible Forest Gardening* in 2005, and Lawton presented the film *Establishing a Food Forest* in 2008.

Austrian Sepp Holzer practices "Holzer Permaculture" on his *Krameterhof* farm, at varying altitudes ranging from 1,100 to 1,500 metres above sea level. His designs create micro-climates with rocks, ponds and living wind barriers, enabling the cultivation of a variety of fruit trees, vegetables and flowers in a region that averages 4 °C, and with temperatures as low as -20 °C in the winter.

Projects

El Pilar on the Belize-Guatemala border features a forest garden to demonstrate traditional Maya agricultural practices. A further 1-acre model forest garden, called Känan K'aax (meaning well-tended garden in Mayan), is being funded by the National Geographic Society and developed at Santa Familia Primary School in Cayo.

In the United States the largest known food forest on public land is believed to be the 7-acre Beacon Food Forest in Seattle, WA. Other forest garden projects include those at the Central Rocky Mountain Permaculture Institute in Basalt, Colorado and Montview Neighborhood farm in Northampton, Massachusetts.

In Canada food forester Richard Walker has been developing and maintaining food forests in the province of British Columbia for over 30 years. He developed a 3-acre food forest that when at maturity provided raw materials for a nursery and herbalism business as well as food for his family. The Living Centre have developed various forest garden projects in Ontario.

In the United Kingdom, other than those run by the Agroforestry Research Trust (ART), there are numerous forest garden projects such as the Bangor Forest Garden in Gwynedd, North West Wales. Martin Crawford from ART administers the Forest Garden Network, an informal network of people and organisations around the world who are cultivating their own forest gardens.

References

- Rosholt, Malcolm (1982). Lumbermen on the Chippewa - The Camp (PDF). Rosholt, Wisconsin: Rosholt House. p. 8. ISBN 0-910417-00-8.
- Nathan, Susan (2005). The Other Side of Israel: My Journey Across the Jewish/Arab Divide. New York: Nan A. Talese. pp. 130–131. ISBN 978-0-385-51456-9.
- Dave Deppner; John Leary; Karin Vermilye; Steve McCrea (2005). The Global Cooling Answer Book (PDF) (Second ed.). Trees for the Future. ISBN 1-879857-20-0. Retrieved 2007.
- David L. Lentz, ed. (2000). Imperfect Balance: Landscape Transformations in the Precolumbian Americas. p. 212. ISBN 9780231111577.
- Khan, Madeline (February 9, 2004). “Clear cut forests in Malaysia offer lessons for logging worldwide.” The Varsity, University of Toronto. Retrieved: 2013-08-06.
- “Preventing Tree Planting Injuries” (PDF). Work Safe BC. Workers’ Compensation Board of British Columbia. 2006. Retrieved December 13, 2013.
- Mellinger, Robert (16 February 2012). “Nation’s Largest Food Forest takes root on Beacon Hill”. Crosscut. Retrieved 14 March 2012.
- Black, R. (6 August 2012). “Gibson settles discord on timber”. BBC News. Archived from the original on 7 August 2012. Retrieved 6 August 2012.
- Vogel, John N. (1982–1983). “The Round Lake logging dam: a survivor of Wisconsin’s log-driving days”. Wisconsin Magazine of History. 66 (3): 176. Retrieved 2011-09-22.
- Toso Bozic (September 14, 2009). “Woodlot Harvest”. Government of Alberta: Agriculture and Rural Development. Retrieved 2011-03-07.
- “Trees from Israel” (PDF). standwithus.com. Archived from the original (PDF) on November 17, 2006. Retrieved 29 October 2011.

4

Ecological Management in Forestry

Forestry was born out of the conservation movement in the late 19th and early 20th century. It has seen various avatars in the hands of professional foresters and has been the source of wealth for many nations. Recent events of climate change, global warming and high concentrations of the urban environment have all led to concerted efforts of sustainable forestry practices. This chapter lists some of them such as landscape ecology and forest dynamics studies.

Soil Management

Soil management concerns all operations, practices, and treatments used to protect soil and enhance its performance.

Environment

According to the EPA, agricultural soil management practices can lead to production and emission of nitrous oxide (N_2O), a major greenhouse gas and air pollutant. Activities that can contribute to N_2O emissions include fertilizer usage, irrigation and tillage. The management of soils accounts for over half of the emissions from the Agriculture sector. Cattle livestocks account for one third of emissions, through methane emissions. Manure management and rice cultivation also emit emissions.

Methods that significantly enhance carbon sequestration in soil include no-till farming, residue mulching, cover cropping, and crop rotation, all of which are more widely used in organic farming than in conventional farming. Because only 5% of US farmland currently uses no-till and residue mulching, there is a large potential for carbon sequestration.

Practices

Soil management practices that affect soil quality:

- Controlling traffic on the soil surface helps to reduce soil compaction, which can reduce aeration and water infiltration.
- Cover crops keep the soil anchored and covered in off-seasons so that the soil is not eroded by wind and rain.
- Crop rotations for row crops alternate high-residue crops with lower-residue crops to increase the amount of plant material left on the surface of the soil during the year to protect the soil from erosion.
- Nutrient management can help to improve the fertility of the soil and the amount of organic matter content, which improves soil structure and function.

- Tillage, especially reduced-tillage or no-till operations limit the amount of soil disturbance while cultivating a new crop, and help to maintain plant residues on the surface of the soil for erosion protection and water retention.

Advantages of Soil Management

- Maintain soil fertility
- Restore soil fertility
- Make the agricultural process an economic one
- Help increase yield

Landscape Ecology

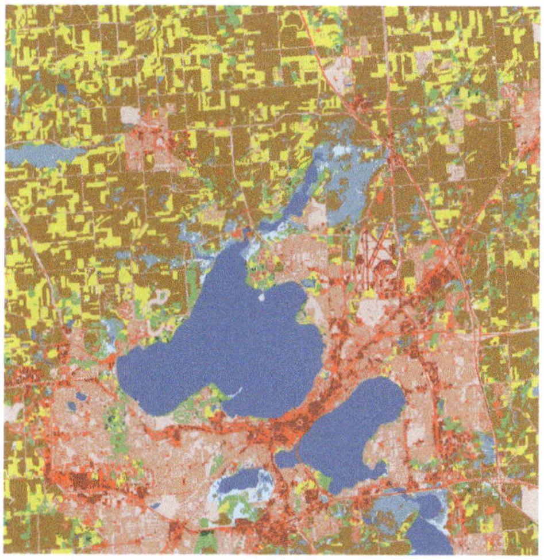

Land cover surrounding Madison, WI. Fields are colored yellow and brown, water is colored blue, and urban surfaces are colored red.

Impervious surfaces surrounding Madison, WI

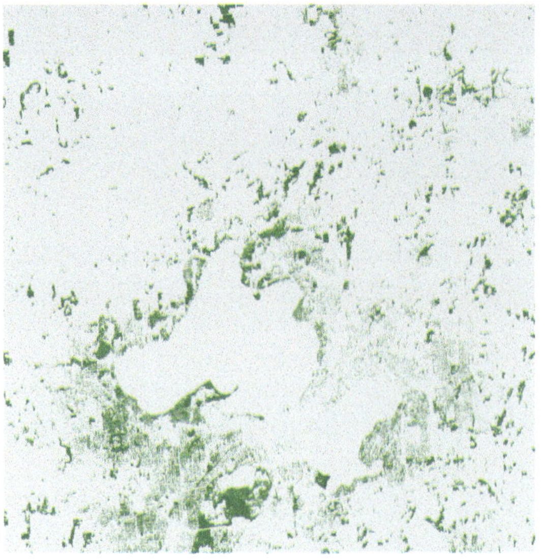

Canopy cover surrounding Madison, WI

Landscape ecology is the science of studying and improving relationships between ecological processes in the environment and particular ecosystems. This is done within a variety of landscape scales, development spatial patterns, and organizational levels of research and policy.

As a highly interdisciplinary field in systems science, landscape ecology integrates biophysical and analytical approaches with humanistic and holistic perspectives across the natural sciences and social sciences. Landscapes are spatially heterogeneous geographic areas characterized by diverse interacting patches or ecosystems, ranging from relatively natural terrestrial and aquatic systems such as forests, grasslands, and lakes to human-dominated environments including agricultural and urban settings. The most salient characteristics of landscape ecology are its emphasis on the relationship among pattern, process and scale, and its focus on broad-scale ecological and environmental issues. These necessitate the coupling between biophysical and socioeconomic sciences. Key research topics in landscape ecology include ecological flows in landscape mosaics, land use and land cover change, scaling, relating landscape pattern analysis with ecological processes, and landscape conservation and sustainability.

Terminology

The German term *Landschaftsökologie*–thus *landscape ecology*–was coined by German geographer Carl Troll in 1939. He developed this terminology and many early concepts of landscape ecology as part of his early work, which consisted of applying aerial photograph interpretation to studies of interactions between environment and vegetation.

Explanation

Heterogeneity is the measure of how parts of a landscape differ from one another. Landscape ecology looks at how this spatial structure affects organism abundance at the landscape level, as well as the behavior and functioning of the landscape as a whole. This includes studying the influence of pattern, or the internal order of a landscape, on process, or the continuous operation of functions of organisms. Landscape ecology also includes geomorphology as applied to the design and architecture of landscapes. Geomorphology is the study of how geological formations are responsible for the structure of a landscape.

History

Evolution of Theory

One central landscape ecology theory originated from MacArthur & Wilson's *The Theory of Island Biogeography*. This work considered the biodiversity on islands as the result of competing forces of colonization from a mainland stock and stochastic extinction. The concepts of island biogeography were generalized from physical islands to abstract patches of habitat by Levins' metapopulation model (which can be applied e.g. to forest islands in the agricultural landscape). This generalization spurred the growth of landscape ecology by providing conservation biologists a new tool to assess how habitat fragmentation affects population viability. Recent growth of landscape ecology owes much to the development of geographic information systems (GIS) and the availability of large-extent habitat data (e.g. remotely sensed datasets).

Development as a Discipline

Landscape ecology developed in Europe from historical planning on human-dominated landscapes. Concepts from general ecology theory were integrated in North America. While general ecology theory and its sub-disciplines focused on the study of more homogenous, discrete community units organized in a hierarchical structure (typically as ecosystems, populations, species, and communities), landscape ecology built upon heterogeneity in space and time. It frequently included human-caused landscape changes in theory and application of concepts.

By 1980, landscape ecology was a discrete, established discipline. It was marked by the organization of the International Association for Landscape Ecology (IALE) in 1982. Landmark book publications defined the scope and goals of the discipline, including Naveh and Lieberman and Forman and Godron. Forman wrote that although study of "the ecology of spatial configuration at the human scale" was barely a decade old, there was strong potential for theory development and application of the conceptual framework. Today, theory and application of landscape ecology continues to develop through a need for innovative applications in a changing landscape and environment. Landscape ecology relies on advanced technologies such as remote sensing, GIS, and models. There has been associated development of powerful quantitative methods to examine the interactions of patterns and processes. An example would be determining the amount of carbon present in the soil based on landform over a landscape, derived from GIS maps, vegetation types, and rainfall data for a region.

Definitions/Conceptions of Landscape Ecology

Nowadays, at least six different conceptions of landscape ecology can be identified: one group tending toward the more disciplinary concept of ecology (subdiscipline of biology; in conceptions 2, 3, and 4) and another group—characterized by the interdisciplinary study of relations between human societies and their environment—inclined toward the integrated view of geography (in conceptions 1, 5, and 6):

1. *Interdisciplinary scientific analysis of subjectively defined landscape units* (e.g. Neef School): Landscapes are defined in terms of uniformity in land use. Landscape ecology explores the landscape's natural potential in terms of functional utility for human societies. To analyse this potential, it is necessary to draw on several natural sciences.

2. *Topological ecology at the landscape scale* (e.g. Forman & Godron): 'Landscape' is defined as a heterogeneous land area composed of a cluster of interacting ecosystems (woods, meadows, marshes, villages, etc.) that is repeated in similar form throughout. It is explicitly stated that landscapes are areas at a kilometres wide "human scale" of perception, modification, etc. Landscape ecology describes and explains the landscapes' characteristic patterns of ecosystems and investigates the flux of energy, mineral nutrients, and species among their component ecosystems, providing important knowledge for addressing land-use issues.

3. *Organism-centred, multi-scale topological ecology* (e.g. John A. Wiens): Explicitly rejecting views expounded by Troll, Zonneveld, Naveh, Forman & Godron, etc., landscape and landscape ecology are defined independently of human perceptions, interests, and modifications of nature. 'Landscape' is defined – regardless of scale – as the 'template' on which spatial patterns influence ecological processes. Not humans, but rather the respective species being studied is the point of reference for what constitutes a landscape.

4. *Topological ecology at the landscape level of biological organisation* (e.g. Urban et al.): On the basis of ecological hierarchy theory, it is presupposed that nature is working at multiple scales and has different levels of organisation which are part of a rate-structured, nested hierarchy. Specifically, it is claimed that, above the ecosystem level, a landscape level exists which is generated and identifiable by high interaction intensity between ecosystems, a specific interaction frequency and, typically, a corresponding spatial scale. Landscape ecology is defined as ecology that focuses on the influence exerted by spatial and temporal patterns on the organisation of, and interaction among, functionally integrated multispecies ecosystems.

5. *Analysis of social-ecological systems using the natural and social sciences and humanities* (e.g. Leser; Naveh; Zonneveld): Landscape ecology is defined as an interdisciplinary super-science that explores the relationship between human societies and their specific environment, making use of not only various natural sciences, but also social sciences and humanities. This conception is grounded in the assumption that social systems are linked to their specific ambient ecological system in such a way that both systems together form a co-evolutionary, self-organising unity called 'landscape'. Societies' cultural, social and economic dimensions are regarded as an integral part of the global ecological hierarchy, and landscapes are claimed to be the manifest systems of the 'Total Human Ecosystem' (Naveh) which encompasses both the physical ('geospheric') and mental ('noospheric') spheres.

6. *Ecology guided by cultural meanings of lifeworldly landscapes* (frequently pursued in practice but not defined): Landscape ecology is defined as ecology that is guided by an external aim, namely, to maintain and develop lifeworldly landscapes. It provides the ecological knowledge necessary to achieve these goals. It investigates how to

sustain and develop those populations and ecosystems which (i) are the material 'vehicles' of lifeworldly, aesthetic and symbolic landscapes and, at the same time, (ii) meet societies' functional requirements, including provisioning, regulating, and supporting ecosystem services. Thus landscape ecology is concerned mainly with the populations and ecosystems which have resulted from traditional, regionally specific forms of land use.

Relationship to Ecological Theory

Some research programmes of landscape ecology theory, namely those standing in the European tradition, may be slightly outside of the "classical and preferred domain of scientific disciplines" because of the large, heterogeneous areas of study. However, general ecology theory is central to landscape ecology theory in many aspects. Landscape ecology consists of four main principles: the development and dynamics of spatial heterogeneity, interactions and exchanges across heterogeneous landscapes, influences of spatial heterogeneity on biotic and abiotic processes, and the management of spatial heterogeneity. The main difference from traditional ecological studies, which frequently assume that systems are spatially homogenous, is the consideration of spatial patterns.

Important Terms in Landscape Ecology

Landscape ecology not only created new terms, but also incorporated existing ecological terms in new ways. Many of the terms used in landscape ecology are as interconnected and interrelated as the discipline itself.

Landscape

Certainly, 'landscape' is a central concept in landscape ecology. It is, however, defined in quite different ways. For example: Carl Troll conceives of landscape not as a mental construct but as an objectively given 'organic entity', a "harmonic individuum of space". Ernst Neef defines landscapes as sections within the uninterrupted earth-wide interconnection of geofactors which are defined as such on the basis of their uniformity in terms of a specific land use, and are thus defined in an anthropocentric and relativistic way. According to Richard Forman and Michael Godron, a landscape is a heterogeneous land area composed of a cluster of interacting ecosystems that is repeated in similar form throughout, whereby they list woods, meadows, marshes and villages as examples of a landscape's ecosystems, and state that a landscape is an area at least a few kilometres wide. John A. Wiens opposes the traditional view expounded by Carl Troll, Isaak S. Zonneveld, Zev Naveh, Richard T. T. Forman/Michel Godron and others that landscapes are arenas in which humans interact with their environments on a kilometre-wide scale; instead, he defines 'landscape'—regardless of scale—as "the template on which spatial patterns influence ecological processes". Some define 'landscape' as an area containing two or more ecosystems in close proximity.

Scale and Heterogeneity (Incorporating Composition, Structure, and Function)

A main concept in landscape ecology is *scale*. Scale represents the real world as translated onto a map, relating distance on a map image and the corresponding distance on earth. Scale is also the spatial or temporal measure of an object or a process, or amount of spatial resolution. Components of scale include composition, structure, and function, which are all important ecological concepts. Applied to landscape ecology, *composition* refers to the number of patch types represented on a landscape and their relative abundance. For example, the amount of forest or wetland, the length of forest edge, or the density of roads can be aspects of landscape composition. *Structure* is determined by the composition, the configuration, and the proportion of different patches across the landscape, while *function* refers to how each element in the landscape interacts based on its life cycle events. *Pattern* is the term for the contents and internal order of a heterogeneous area of land.

A landscape with structure and pattern implies that it has spatial *heterogeneity*, or the uneven distribution of objects across the landscape. Heterogeneity is a key element of landscape ecology that separates this discipline from other branches of ecology. Landscape heterogeneity is able to quantify with agent-based methods as well.

Patch and Mosaic

Patch, a term fundamental to landscape ecology, is defined as a relatively homogeneous area that differs from its surroundings. Patches are the basic unit of the landscape that change and fluctuate, a process called *patch dynamics*. Patches have a definite shape and spatial configuration, and can be described compositionally by internal variables such as number of trees, number of tree species, height of trees, or other similar measurements.

Matrix is the "background ecological system" of a landscape with a high degree of connectivity. *Connectivity* is the measure of how connected or spatially continuous a corridor, network, or matrix is. For example, a forested landscape (matrix) with fewer gaps in forest cover (open patches) will have higher connectivity. *Corridors* have important functions as strips of a particular type of landscape differing from adjacent land on both sides. A *network* is an interconnected system of corridors while *mosaic* describes the pattern of patches, corridors, and matrix that form a landscape in its entirety.

Boundary and Edge

Landscape patches have a boundary between them which can be defined or fuzzy. The zone composed of the edges of adjacent ecosystems is the *boundary*. *Edge* means the portion of an ecosystem near its perimeter, where influences of the adjacent patches can cause an environmental difference between the interior of the patch and its edge. This edge effect includes a distinctive species composition or abundance. For example, when a landscape is a mosaic of perceptibly different types, such as a forest adjacent to a grassland, the edge is the location where the two types adjoin. In a continuous landscape, such as a forest giving way to open woodland, the exact edge location is fuzzy and is sometimes determined by a local gradient exceeding a threshold, such as the point where the tree cover falls below thirty-five percent.

Ecotones, Ecoclines, and Ecotopes

A type of boundary is the *ecotone*, or the transitional zone between two communities. Ecotones can arise naturally, such as a lakeshore, or can be human-created, such as a cleared agricultural field from a forest. The ecotonal community retains characteristics of each bordering community and often contains species not found in the adjacent communities. Classic examples of ecotones include fencerows, forest to marshlands transitions, forest to grassland transitions, or land-water interfaces such as riparian zones in forests. Characteristics of ecotones include vegetational sharpness, physiognomic change, occurrence of a spatial community mosaic, many exotic species, ecotonal species, spatial mass effect, and species richness higher or lower than either side of the ecotone.

An *ecocline* is another type of landscape boundary, but it is a gradual and continuous change in environmental conditions of an ecosystem or community. Ecoclines help explain the distribution and diversity of organisms within a landscape because certain organisms survive better under certain conditions, which change along the ecocline. They contain heterogeneous communities which are considered more environmentally stable than those of ecotones.

An *ecotope* is a spatial term representing the smallest ecologically distinct unit in mapping and classification of landscapes. Relatively homogeneous, they are spatially explicit landscape units used to stratify landscapes into ecologically distinct features. They are useful for the measurement and mapping of landscape structure, function, and change over time, and to examine the effects of disturbance and fragmentation.

Disturbance and Fragmentation

Disturbance is an event that significantly alters the pattern of variation in the structure or function of a system. *Fragmentation* is the breaking up of a habitat, ecosystem, or land-use type into smaller parcels. Disturbance is generally considered a natural process. Fragmentation causes land transformation, an important process in landscapes as development occurs.

An important consequence of repeated, random clearing (whether by natural disturbance or human activity) is that contiguous cover can break down into isolated patches. This happens when the area cleared exceed a critical level, which means that landscapes exhibit two phases: connected and disconnected.

Landscape Ecology Theory

Landscape ecology theory stresses the role of human impacts on landscape structures and functions. It also proposes ways for restoring degraded landscapes. Landscape ecology explicitly includes humans as entities that cause functional changes on the landscape. Landscape ecology theory includes the landscape stability principle, which emphasizes the importance of landscape structural heterogeneity in developing resistance to disturbances, recovery from disturbances, and promoting total system stability. This principle is a major contribution to general ecological theories which highlight the importance of relationships among the various components of the landscape. Integrity of landscape components helps maintain resistance to external threats, including development and land transformation by human activity. Analysis of land use change has included a strongly geographical approach which has led to the acceptance of the idea of multifunctional properties of landscapes. There are still calls for a more unified theory of landscape ecology due to differences in professional opinion among ecologists and its interdisciplinary approach (Bastian 2001).

An important related theory is hierarchy theory, which refers to how systems of discrete functional elements operate when linked at two or more scales. For example, a forested landscape might be hierarchically composed of drainage basins, which in turn are composed of local ecosystems, which are in turn composed of individual trees and gaps. Recent theoretical developments in landscape ecology have emphasized the relationship between pattern and process, as well as the effect that changes in spatial scale has on the potential to extrapolate information across scales. Several studies suggest that the landscape has critical thresholds at which ecological processes will show dramatic changes, such as the complete transformation of a landscape by an invasive species due to small changes in temperature characteristics which favor the invasive's habitat requirements.

Landscape Ecology Application

Research Directions

Developments in landscape ecology illustrate the important relationships between spatial patterns and ecological processes. These developments incorporate quantitative methods that link spatial patterns and ecological processes at broad spatial and temporal scales. This linkage of time, space, and environmental change can assist managers in applying plans to solve environmental problems. The increased attention in recent years on spatial dynamics has highlighted the need for new quantitative methods that can analyze patterns, determine the importance of spatially explicit processes, and develop reliable models. Multivariate analysis techniques are frequently used to examine landscape level vegetation patterns. Studies use statistical techniques, such as cluster analysis, canonical correspondence analysis (CCA), or detrended correspondence analysis (DCA), for classifying vegetation. Gradient analysis is another way to determine the vegetation structure across a landscape or to help delineate critical wetland habitat for conservation or mitigation purposes (Choesin and Boerner 2002).

Climate change is another major component in structuring current research in landscape ecology. Ecotones, as a basic unit in landscape studies, may have significance for management under climate change scenarios, since change effects are likely to be seen at ecotones first because of the unstable nature of a fringe habitat. Research in northern regions has examined landscape ecological processes, such as the accumulation of snow, melting, freeze-thaw action, percolation, soil moisture variation, and temperature regimes through long-term measurements in Norway. The study analyzes gradients across space and time between ecosystems of the central high mountains to determine relationships between distribution patterns of animals in their environment. Looking at where animals live, and how vegetation shifts over time, may provide insight into changes in snow and ice over long periods of time across the landscape as a whole.

Other landscape-scale studies maintain that human impact is likely the main determinant of landscape pattern over much of the globe. Landscapes may become substitutes for biodiversity measures because plant and animal composition differs between samples taken from sites within different landscape categories. Taxa, or different species, can "leak" from one habitat into another, which has implications for landscape ecology. As human land use practices expand and continue to increase the proportion of edges in landscapes, the effects of this leakage across edges on assemblage integrity may become more significant in conservation. This is because taxa may be conserved across landscape levels, if not at local levels.

Relationship to Other Disciplines

Landscape ecology has been incorporated into a variety of ecological subdisciplines. For example, a recent development has been the more explicit consideration of spatial concepts and principles applied to the study of lakes, streams, and wetlands in the field of landscape limnology. In addition, landscape ecology has important links to application-oriented disciplines such as agriculture and forestry. In agriculture, landscape ecology has introduced new options for the management of environmental threats brought about by the intensification of agricultural practices. Agriculture has always been a strong human impact on ecosystems. In forestry, from structuring stands for fuelwood and timber to ordering stands across landscapes to enhance aesthetics, consumer needs have affected conservation and use of forested landscapes. Landscape forestry provides methods, concepts, and analytic procedures for landscape forestry. Finally, landscape ecology has been cited as a contributor to the development of fisheries biology as a distinct biological science discipline, and is frequently incorporated in study design for wetland delineation in hydrology. It has helped shape integrated landscape management.

Forest Ecology

The Daintree Rainforest in Queensland, Australia

Forest ecology is the scientific study of the interrelated patterns, processes, flora, fauna and ecosystems in forests. The management of forests is known as forestry, silviculture, and forest management. A forest ecosystem is a natural woodland unit consisting of all plants, animals and micro-organisms (Biotic components) in that area functioning together with all of the non-living physical (abiotic) factors of the environment. The forest ecosystem is very important.

Relationship to Other Branches of Ecology

Forest ecology is one branch of a biotically-oriented classification of types of ecological study (as opposed to a classification based on organizational level or complexity, for example population or community ecology). Thus, forests are studied at a number of organizational levels, from the individual organism to the ecosystem. However, as the term forest connotes an area inhabited by more than one organism, forest ecology most often concentrates on the level of the population, community or ecosystem. Logically, trees are an important component of forest research, but the wide variety of other life forms and abiotic components in most forests means that other elements, such as wildlife or soil nutrients, are often the focal point. Thus, forest ecology is a highly diverse and important branch of ecological study.

Forest ecology studies share characteristics and methodological approaches with other areas of terrestrial plant ecology. However, the presence of trees makes forest ecosystems and their study unique in numerous ways.

Community Diversity and Complexity

Since trees can grow larger than other plant life-forms, there is the potential for a wide variety of forest structures (or physiognomies). The infinite number of possible spatial arrangements of trees of varying size and species makes for a highly intricate and diverse micro-environment in which environmental variables such as solar radiation, temperature, relative humidity, and wind speed can vary considerably over large and small distances. In addition, an important proportion of a forest ecosystem's biomass is often underground, where soil structure, water quality and quantity, and levels of various soil nutrients can vary greatly. Thus, forests are often highly heterogeneous environments compared to other terrestrial plant communities. This heterogeneity in turn can enable great biodiversity of species of both plants and animals. It also affects the design of forest inventory sampling strategies, the results of which are sometimes used in ecological studies. A number of factors within the forest affect biodiversity; primary factors enhancing wildlife abundance and biodiversity are the presence of diverse tree species within the forest and the absence of even aged timber management. For example, the wild turkey thrives when uneven heights and canopy variations exist and its numbers are diminished by even aged timber management.

Energy Flux

Forest ecologists are interested in the effects of large disturbances, such as wildfires. Montana, USA.

Forests accumulate large amounts of standing biomass, and many are capable of accumulating it at high rates, i.e. they are highly productive. Such high levels of biomass and tall vertical structures represent large stores of potential energy that can be converted to kinetic energy under the right circumstances. Two such conversions of great importance are fires and treefalls, both of which radically alter the biota and the physical environment where they occur. Also, in forests of high productivity, the rapid growth of the trees themselves induces biotic and environmental changes, although at a slower rate and lower intensity than relatively instantaneous disturbances such as fires.

Death and Regeneration

Woody material, often referred to as coarse woody debris, decays relatively slowly in many forests in comparison to most other organic materials, due to a combination of environmental factors and wood chemistry. Trees growing in arid and/or cold environments do so especially slowly. Thus, tree trunks and branches can remain on the forest floor for long periods, affecting such things as wildlife habitat, fire behavior, and tree regeneration processes.

Water

Lastly, forest trees store large amounts of water because of their large size and anatomical/physiological characteristics. They are therefore important regulators of hydrological processes, especially those involving groundwater hydrology and local evaporation and rainfall/snowfall patterns. Thus, forest ecological studies are sometimes closely aligned with meteorological and hydrological studies in regional ecosystem or resource planning studies. Perhaps more importantly the duff or leaf litter can form a major repository of water storage. When this litter is removed or compacted (e.g. through grazing or human overuse), erosion and flooding are exacerbated as well as deprivation of dry season water for forest organisms.

Ecological Potential of Forestal Species

The ecological potential of a particular species is a measure of its capacity to effectively compete in a given geographical area, ahead of other species, as they all try to occupy a natural space. For some areas it has been quantified, as for instance by Hans-Jürgen Otto, for central Europe. He takes three groups of parameters:

- Related to site requirements: Tolerance to low temperatures, tolerance to dry climate, frugality.
- Specific qualities: Shade tolerance, height growth, stability, longevity, regeneration capacity.
- Specific risks: Resistance to late freezing, resistance to wind/ice storm, resistance to fire, resistance to biotic agents.

Every parameter is scored between 0 and 5 for each considered species, and then a global mean value calculated. A value above 3.5 is considered high, below 3.0 low, and intermediate for those in between. In this study *Fagus sylvatica* has a score of 3.82, *Fraxinus excelsior* 3.08 and *Juglans regia* 2.92; and are examples of the three categories.

Forest Dynamics

Forest dynamics describes the underlying physical and biological forces that shape and change a forest ecosystem. The continuous state of change in forests can be summarized with two basic elements: disturbance and succession.

Disturbance

Forest disturbances are events that cause change in the structure and composition of a forest ecosystem, beyond the growth and death of individual organisms. Disturbances can vary in frequency and intensity, and include natural disasters such as fire, landslides, wind, and volcanic eruptions, outbreaks of insects, fungi, and other pathogens, animal-caused effects such as grazing and trampling, and anthropogenic disturbances such as logging, pollution, the clearing of land for urbanization or agriculture, and the introduction of invasive species. Not all disturbances are destructive or negative to the overall forest ecosystem. Many natural disturbances allow for renewal and growth and often release necessary nutrients.

Small-scale disturbances are the key to creating and maintaining diversity and heterogeneity within a forest. Small-scale disturbances are events such as single-tree blowdowns, which create gaps that let light through the canopy to the understory and forest floor. This available light allows early-successional shade-intolerant species to colonize and maintain a population within the dominant forest, leading to the complex spatial mosaic forest structure recognized as old-growth. This process is referred to as patch dynamics or gap dynamics and has been described across many types of forests, including tropical, temperate, and boreal.

The sets and patterns of natural disturbances that characterize a particular area or ecosystem are referred to as the ecosystem's *disturbance regime*. A natural community is closely linked with its natural disturbance regime. For example, temperate and boreal rainforests typically have a disturbance regime consisting of high-frequency but small-scale events, resulting in a highly complex forest dominated by very old trees. In contrast, forests that have a disturbance regime consisting of high-severity stand-replacing events, such as frequent fires, tend to be more uniform in structure and have relatively young tree stands.

Succession

Forest succession is the process by which species recover and regenerate after a disturbance. The type of disturbance, the climate and weather conditions, the presence of colonizing species, and the interactions among species all influence the path that succession will take. Species diversity and composition fluctuate throughout succession. The classic model of succession is known as *relay floristics* and refers to a relay of dominant species. After a stand-replacing disturbance, shade-intolerant species colonize and grow into a dominant canopy, but due to their shade-intolerance they are unable to regenerate under their own canopy; the understory (composed of shade-tolerant species) gradually replaces the canopy, and due to its shade-tolerance it can regenerate under its own canopy and therefore becomes the dominant species. Often succession is not so complete or directed as the relay floristics model describes. Species can be mid-tolerant of shade and survive by taking advantage of small amounts of light coming through the canopy, and further disturbances can create small gaps. These and other factors can lead to a mixture of dominant species and a not so obvious "end" to succession (climax community).

Many successional trajectories follow a basic four-stage development pattern. The first of these stages, *stand initiation*, occurs after a major disturbance and involves many species arriving in the area of abundant light and nutrients. The second stage, *stem exclusion*, describes the growth and competition of these species as resources become less available; likely one or a few species outcompetes and becomes stand-dominating. The third stage, *understory reinitiation*, involves further disturbance and the creation of gaps; at this point stratification develops, with layers of canopy, midstory, and understory appearing. The final stage, known as *old-growth*, is the extension and completion of the understory reinititation; a complex multi-aged and multi-layered forest has developed.

Considering Climate Change

Forests are sensitive to climate, and so climate change can have a great effect on the dynamics of the ecosystem. Rising carbon dioxide levels can increase the productivity and growth of trees, which will then decrease as other nutrients become limiting. Changes in temperature and precipitation can affect the success of various species and the resulting species assemblage. Many factors of climate change can also affect an ecosystem's disturbance regime, making the forest more or less susceptible to different disturbances and altering or even preventing recovery after a disturbance.

Importance

Forests offer many ecosystem services including timber, fresh water, carbon storage, and areas of recreation. To conserve these services, along with the natural habitat and biodiversity that forests provide, understanding the dynamics that are creating and maintaining the forests is a priority. Forestry and silviculture operations require a thorough comprehension of forest dynamics in order to implement effective management and conservation techniques.

Ecological Succession

Succession after disturbance: a boreal forest one year (left) and two years (right) after a wildfire.

Ecological succession is the process of change in the species structure of an ecological community over time. The time scale can be decades (for example, after a wildfire), or even millions of years after a mass extinction.

The community begins with relatively few pioneering plants and animals and develops through increasing complexity until it becomes stable or self-perpetuating as a climax community. The engine of succession, the cause of ecosystem change, is the impact of established species upon their own environments. A consequence of living is the sometimes subtle and sometimes overt alteration of one's own environment.

It is a phenomenon or process by which an ecological community undergoes more or less orderly and predictable changes following a disturbance or the initial colonization of a new habitat. Succession may be initiated either by formation of new, unoccupied habitat, such as from a lava flow or a severe landslide, or by some form of disturbance of a community, such as from a fire, severe windthrow, or logging. Succession that begins in new habitats, uninfluenced by pre-existing communities is called primary succession, whereas succession that follows disruption of a pre-existing community is called secondary succession.

Succession was among the first theories advanced in ecology. The study of succession remains at the core of ecological science. Ecological succession was first documented in the Indiana Dunes of Northwest Indiana which led to efforts to preserve the Indiana Dunes. Exhibits on ecological succession are displayed in the Hour Glass, a museum in Ogden Dunes.

History

Precursors of the idea of ecological succession go back to the beginning of the 19th century. The French naturalist Adolphe Dureau de la Malle was the first to make use of the word *succession* concerning the vegetation development after forest clear-cutting. In 1859 Henry David Thoreau wrote an address called "The Succession of Forest Trees" in which he described succession in an oak-pine forest. "It has long been known to observers that squirrels bury nuts in the ground, but I am not aware that any one has thus accounted for the regular succession of forests." The Austrian botanist Anton Kerner published a study about the succession of plants in the Danube river basin in 1863.

H. C. Cowles

Henry Chandler Cowles, at the University of Chicago, developed a more formal concept of succession. Inspired by studies of Danish dunes by Eugen Warming, Cowles studied vegetation development on sand dunes on the shores of Lake Michigan (the Indiana Dunes). He recognized that vegetation on dunes of different ages might be interpreted as different stages of a general trend of vegetation development on dunes (an approach to the study of vegetation change later termed space-for-time substitution, or chronosequence studies). He first published this work as a paper in the *Botanical Gazette* in 1899 ("The ecological relations of the vegetation of the sand dunes of Lake Michigan"). In this classic publication and subsequent papers, he formulated the idea of primary succession and the notion of a sere—a repeatable sequence of community changes specific to particular environmental circumstances.

The Indiana Dunes on Lake Michigan, which stimulated Cowles' development of his theories of ecological succession

Gleason and Clements

From about 1900 to 1960, however, understanding of succession was dominated by the theories of Frederic Clements, a contemporary of Cowles, who held that seres were highly predictable and deterministic and converged on a climatically determined stable climax community regardless of starting conditions. Clements explicitly analogized the successional development of ecological communities with ontogenetic development of individual organisms, and his model is often referred to as the pseudo-organismic theory of community ecology. Clements and his followers developed a complex taxonomy of communities and successional pathways.

Henry Gleason offered a contrasting framework as early as the 1920s. The Gleasonian model was more complex and much less deterministic than the Clementsian. It differs most fundamentally from the Clementsian view in suggesting a much greater role of chance factors and in denying the existence of coherent, sharply bounded community types. Gleason argued that species distributions responded individualistically to environmental factors, and communities were best regarded as artifacts of the juxtaposition of species distributions. Gleason's ideas, first published in 1926, were largely ignored until the late 1950s.

Two quotes illustrate the contrasting views of Clements and Gleason. Clements wrote in 1916:

The developmental study of vegetation necessarily rests upon the assumption that the unit or climax formation is an organic entity. As an organism the formation arises, grows, matures, and dies. Furthermore, each climax formation is able to reproduce itself, repeating with essential fidelity the stages of its development.

—Frederic Clements

while Gleason, in his 1926 paper, said:

An association is not an organism, scarcely even a vegetational unit, but merely a coincidence.

—Henry Gleason

Gleason's ideas were, in fact, more consistent with Cowles' original thinking about succession. About Clements' distinction between primary succession and secondary succession, Cowles wrote (1911):

This classification seems not to be of fundamental value, since it separates such closely related phenomena as those of erosion and deposition, and it places together such unlike things as human agencies and the subsidence of land.

—*Henry Cowles*

Modern Era

A more rigorous, data-driven testing of successional models and community theory generally began with the work of Robert Whittaker and John Curtis in the 1950s and 1960s. Succession theory has since become less monolithic and more complex. J. Connell and R. Slatyer attempted a codification of successional processes by mechanism. Among British and North American ecologists, the notion of a stable climax vegetation has been largely abandoned, and successional processes have come to be seen as much less deterministic, with important roles for historical contingency and for alternate pathways in the actual development of communities. Debates continue as to the general predictability of successional dynamics and the relative importance of equilibrial vs. non-equilibrial processes. Former Harvard professor F. A. Bazzaz introduced the notion of *scale* into the discussion, as he considered that at local or small area scale the processes are stochastic and patchy, but taking bigger regional areas into consideration, certain tendencies can not be denied.

Factors

The trajectory of successional change can be influenced by site conditions, by the character of the events initiating succession (perturbations), by the interactions of the species present, and by more stochastic factors such as availability of colonists or seeds or weather conditions at the time of disturbance. Some of these factors contribute to predictability of succession dynamics; others add more probabilistic elements. Two important perturbation factors today are human actions and climatic change.

In general, communities in early succession will be dominated by fast-growing, well-dispersed species (opportunist, fugitive, or r-selected life-histories). As succession proceeds, these species will tend to be replaced by more competitive (k-selected) species.

Trends in ecosystem and community properties in succession have been suggested, but few appear to be general. For example, species diversity almost necessarily increases during early succession as new species arrive, but may decline in later succession as competition eliminates opportunistic species and leads to dominance by locally superior competitors. Net Primary Productivity, biomass, and trophic properties all show variable patterns over succession, depending on the particular system and site.

Ecological succession was formerly seen as having a stable end-stage called the climax, sometimes referred to as the 'potential vegetation' of a site, and shaped primarily by the local climate. This idea has been largely abandoned by modern ecologists in favor of nonequilibrium ideas of ecosystems dynamics. Most natural ecosystems experience disturbance at a rate that makes a "climax" community unattainable. Climate change often occurs at a rate and frequency sufficient to prevent arrival at a climax state. Additions to available species pools through range expansions and introductions can also continually reshape communities.

The development of some ecosystem attributes, such as soil properties and nutrient cycles, are both influenced by community properties, and, in turn, influence further successional development. This feed-back process may occur only over centuries or millennia. Coupled with the stochastic nature of disturbance events and other long-term (e.g., climatic) changes, such dynamics make it doubtful whether the 'climax' concept ever applies or is particularly useful in considering actual vegetation.

Types

Primary, Secondary and Cyclic Succession

An example of Secondary Succession by stages:
1. A stable deciduous forest community
2. A disturbance, such as a wild fire, destroys the forest
3. The fire burns the forest to the ground
4. The fire leaves behind empty, but not destroyed, soil
5. Grasses and other herbaceous plants grow back first
6. Small bushes and trees begin to colonize the area
7. Fast growing evergreen trees develop to their fullest, while shade-tolerant trees develop in the understory
8. The short-lived and shade intolerant evergreen trees die as the larger deciduous trees overtop them. The ecosystem is now back to a similar state to where it began.

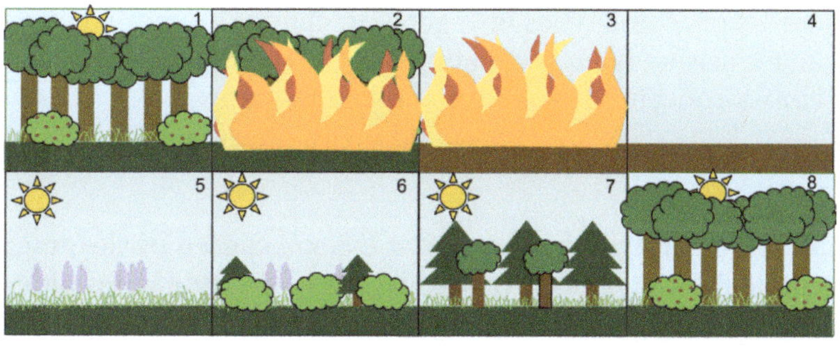

Main articles: Primary succession, Secondary succession, and Cyclic succession

Successional dynamics beginning with colonization of an area that has not been previously occupied by an ecological community, such as newly exposed rock or sand surfaces, lava flows, newly exposed glacial tills, etc., are referred to as primary succession. The stages of primary succession include pioneer plants (lichens and mosses), grassy stage, smaller shrubs, and trees. Animals begin to return when there is food there for them to eat. When it is a fully functioning ecosystem, it has reached the climax community stage. For example, parts of Acadia National Park in Maine went through primary succession.

Successional dynamics following severe disturbance or removal of a pre-existing community are called secondary succession. Dynamics in secondary succession are strongly influenced by pre-disturbance conditions, including soil development, seed banks, remaining organic matter, and residual living organisms. Because of residual fertility and pre-existing organisms, community change in early stages of secondary succession can be relatively rapid. In a fragmented old field habitat created in eastern Kansas, woody plants "colonized more rapidly (per unit area) on large and near-by patches."

Secondary succession: trees are colonizing uncultivated fields and meadows.

Secondary succession is much more commonly observed and studied than primary succession. Particularly common types of secondary succession include responses to natural disturbances such as fire, flood, and severe winds, and to human-caused disturbances such as logging and agriculture. As an example, secondary succession has been occurring in Shenandoah National Park following the 1995 flood of the Mormon River, which destroyed plant and animal life. Today, plant and animal species are beginning to return.

Seasonal and Cyclic Dynamics

Unlike secondary succession, these types of vegetation change are not dependent on disturbance but are periodic changes arising from fluctuating species interactions or recurring events. These models modify the climax concept towards one of dynamic states.

Causes of Plant Succession

Autogenic succession can be brought by changes in the soil caused by the organisms there. These changes include accumulation of organic matter in litter or humic layer, alteration of soil nutrients, change in pH of soil by plants growing there. The structure of the plants themselves can also alter the community. For example, when larger species like trees mature, they produce shade on to the developing forest floor that tends to exclude light-requiring species. Shade-tolerant species will invade the area.

Allogenic succession is caused by external environmental influences and not by the vegetation. For example, soil changes due to erosion, leaching or the deposition of silt and clays can alter the nutrient content and water relationships in the ecosystems. Animals also play an important role in allogenic changes as they are pollinators, seed dispersers and herbivores. They can also increase nutrient content of the soil in certain areas, or shift soil about (as termites, ants, and moles do) creating patches in the habitat. This may create regeneration sites that favor certain species.

Climatic factors may be very important, but on a much longer time-scale than any other. Changes in temperature and rainfall patterns will promote changes in communities. As the climate warmed at the end of each ice age, great successional changes took place. The tundra vegetation and bare glacial till deposits underwent succession to mixed deciduous forest. The greenhouse effect resulting in increase in temperature is likely to bring profound Allogenic changes in the next century. Geological and climatic catastrophes such as volcanic eruptions, earthquakes, avalanches, meteors, floods, fires, and high wind also bring allogenic changes.

Mechanisms

In 1916, Frederic Clements published a descriptive theory of succession and advanced it as a general ecological concept. His theory of succession had a powerful influence on ecological thought. Clements' concept is usually termed classical ecological theory. According to Clements, succession is a process involving several phases:

1. Nudation: Succession begins with the development of a bare site, called Nudation (disturbance).
2. Migration: It refers to arrival of propagules.
3. Ecesis: It involves establishment and initial growth of vegetation.
4. Competition: As vegetation becomes well established, grow, and spread, various species begin to compete for space, light and nutrients.
5. Reaction: During this phase autogenic changes such as the buildup of humus affect the habitat, and one plant community replaces another.
6. Stabilization: A supposedly stable climax community forms.

Seral Communities

A hydrosere community

A seral community is an intermediate stage found in an ecosystem advancing towards its climax community. In many cases more than one seral stage evolves until climax conditions are attained. A *prisere* is a collection of seres making up the development of an area from non-vegetated surfaces to a climax community. Depending on the substratum and climate, different seres are found.

Changes in Animal Life

Succession theory was developed primarily by botanists. The study of succession applied to whole ecosystems initiated in the writings of Ramon Margalef, while Eugene Odum's publication of *The Strategy of Ecosystem Development* is considered its formal starting point.

Animal life also exhibit changes with changing communities. In lichen stage the fauna is sparse. It comprises few mites, ants and spiders living in the cracks and crevices. The fauna undergoes a qualitative increase during herb grass stage. The animals found during this stage include nematodes, insects larvae, ants, spiders, mites, etc. The animal population increases and diversifies with the development of forest climax community. The fauna consists of invertebrates like slugs, snails, worms, millipedes, centipedes, ants, bugs; and vertebrates such as squirrels, foxes, mice, moles, snakes, various birds, salamanders and frogs.

Microsuccession

Succession of micro-organisms including fungi and bacteria occurring within a microhabitat is known as microsuccession or serule. This type of succession occurs in recently disturbed communities or newly available habitat, for example in recently dead trees, animal droppings, exposed glacial till, etc. Microbial communities may also change due to products secreted by the bacteria present. Changes of pH in a habitat could provide ideal conditions for a new species to inhabit the area. In some cases the new species may outcompete the present ones for nutrients leading to the primary species demise. Changes can also occur by microbial succession with variations in water availability and temperature. Theories of macroecology have only recently been applied to microbiology and so much remains to be understood about this growing field. A recent study of microbial succession evaluated the balances between stochastic and deterministic processes in the bacterial colonization of a salt marsh chronosequence. The results of this study show that, much like in macro succession, early colonization (primary succession) is mostly influenced by stochasticity while secondary succession of these bacterial communities was more strongly influenced by deterministic factors.

Climax Concept

According to classical ecological theory, succession stops when the sere has arrived at an equilibrium or steady state with the physical and biotic environment. Barring major disturbances, it will persist indefinitely. This end point of succession is called climax.

Climax Community

The final or stable community in a sere is the *climax community* or *climatic vegetation*. It is self-perpetuating and in equilibrium with the physical habitat. There is no net annual accumulation of organic matter in a climax community. The annual production and use of energy is balanced in such a community.

Characteristics

- The vegetation is tolerant of environmental conditions.

- It has a wide diversity of species, a well-drained spatial structure, and complex food chains.
- The climax ecosystem is balanced. There is equilibrium between gross primary production and total respiration, between energy used from sunlight and energy released by decomposition, between uptake of nutrients from the soil and the return of nutrient by litter fall to the soil.
- Individuals in the climax stage are replaced by others of the same kind. Thus the species composition maintains equilibrium.
- It is an index of the climate of the area. The life or growth forms indicate the climatic type.

Types of Climax

Climatic Climax

If there is only a single climax and the development of climax community is controlled by the climate of the region, it is termed as climatic climax. For example, development of Maple-beech climax community over moist soil. Climatic climax is theoretical and develops where physical conditions of the substrate are not so extreme as to modify the effects of the prevailing regional climate.

Edaphic Climax

When there are more than one climax communities in the region, modified by local conditions of the substrate such as soil moisture, soil nutrients, topography, slope exposure, fire, and animal activity, it is called *edaphic climax*. Succession ends in an edaphic climax where topography, soil, water, fire, or other disturbances are such that a climatic climax cannot develop.

Catastrophic Climax

Climax vegetation vulnerable to a catastrophic event such as a wildfire. For example, in California, chaparral vegetation is the final vegetation. The wildfire removes the mature vegetation and decomposers. A rapid development of herbaceous vegetation follows until the shrub dominance is re-established. This is known as catastrophic climax.

Disclimax

When a stable community, which is not the climatic or edaphic climax for the given site, is maintained by man or his domestic animals, it is designated as Disclimax (disturbance climax) or anthropogenic subclimax (man-generated). For example, overgrazing by stock may produce a desert community of bushes and cacti where the local climate actually would allow grassland to maintain itself.

Subclimax

The prolonged stage in succession just preceding the climatic climax is *subclimax*.

Preclimax and Postclimax

In certain areas different climax communities develop under similar climatic conditions. If the community has life forms lower than those in the expected climatic climax, it is called *preclimax*; a community that has life forms higher than those in the expected climatic climax is *postclimax*. Preclimax strips develop in less moist and hotter areas, whereas Postclimax strands develop in more moist and cooler areas than that of surrounding climate.

Theories

There are three schools of interpretations explaining the climax concept:

- Monoclimax or Climatic Climax Theory was advanced by Clements (1916) and recognizes only one climax whose characteristics are determined solely by climate (climatic climax). The processes of succession and modification of environment overcome the effects of differences in topography, parent material of the soil, and other factors. The whole area would be covered with uniform plant community. Communities other than the climax are related to it, and are recognized as subclimax, postclimax and disclimax.
- Polyclimax Theory was advanced by Tansley (1935). It proposes that the climax vegetation of a region consists of more than one vegetation climaxes controlled by soil moisture, soil nutrients, topography, slope exposure, fire, and animal activity.
- Climax Pattern Theory was proposed by Whittaker (1953). The climax pattern theory recognizes a variety of climaxes governed by responses of species populations to biotic and abiotic conditions. According to this theory the total environment of the ecosystem determines the composition, species structure, and balance of a climax community. The environment includes the species responses to moisture, temperature, and nutrients, their biotic relationships, availability of flora and fauna to colonize the area, chance dispersal of seeds and animals, soils, climate, and disturbance such as fire and wind. The nature of climax vegetation will change as the environment changes. The climax community represents a pattern of populations that corresponds to and changes with the pattern of environment. The central and most widespread community is the climatic climax.

More recently another possible idea has been put forward called the theory of alternative stable states which suggests that there is not one end point but many which transition between each other over ecological time.

Forest Succession

The forests, being an ecological system, are subject to the species succession process. There are "opportunistic" or "pioneer" species that produce great quantities of seed that are disseminated by the wind, and therefore can colonize big empty extensions. They are capable of germinating and growing in direct sunlight. Once they have produced a *closed canopy*, the lack of direct sun radiation at soil makes it difficult for their own seedlings to develop. It is then the opportunity for shade-tolerant species to become established under the protection of the pioneers. When the pioneers die, the shade-tolerant species replace them. These species are capable of growing beneath the canopy, and therefore, in the absence of catastrophes, will stay. For this reason it is then said the stand has reached its climax. When a catastrophe occurs, the opportunity for the pioneers opens up again, provided they are present or within a reasonable range.

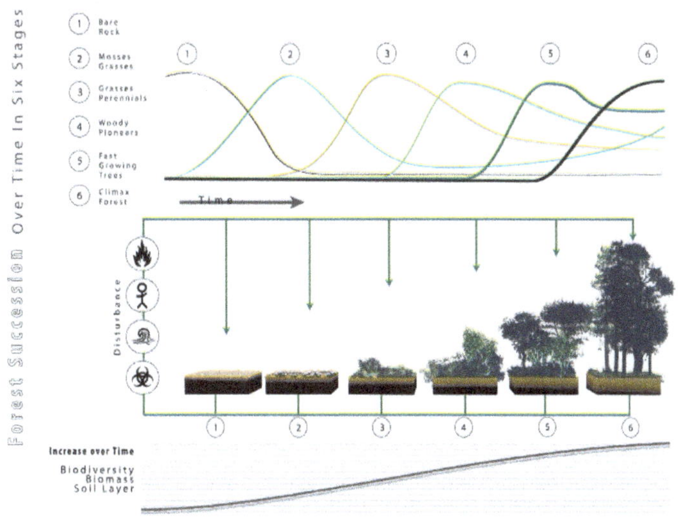

An example of pioneer species, in forests of northeastern North America are *Betula papyrifera* (White birch) and *Prunus serotina* (Black cherry), that are particularly well-adapted to exploit large gaps in forest canopies, but are intolerant of shade and are eventually replaced by other shade-tolerant species in the absence of disturbances that create such gaps.

Things in nature are not black and white, and there are intermediate stages. It is therefore normal that between the two extremes of light and shade there is a gradient, and there are species that may act as pioneer or tolerant, depending on the circumstances. It is of paramount importance to know the tolerance of species in order to practice an effective silviculture.

References

- Otto, Hans-Jürgen (1998). Écologie Forestière (in French). Paris: Institut pour le Développement Forestier. ISBN 9782904740657.
- Michael G. Barbour and William Dwight Billings (2000) North American Terrestrial Vegetation, Cambridge University Press, 708 pages ISBN 0-521-55986-3, ISBN 978-0-521-55986-7
- Kotar, John. "Forest Dynamics". Approaches to Ecological Based Forest Management on Private Lands. USFS and University of Minnesota Extension Service. Retrieved 26 November 2014.
- Baker, Jim; Hunter, Charles (8 December 2013). "Ecological Basis of Silviculture". Southern Forest Resource Assessment. Retrieved 26 November 2014.
- Stone, Emily; Menendez, Lydia (20 October 2011). "Natural Disturbance Regime". The Encyclopedia of Earth. Retrieved 26 November 2014.
- "Secondary succession in an experimentally fragmented landscape: Community patterns across space and time". The U.S. Department of Agriculture. Retrieved 2013-09-30.

5

Forest Resource Management and Forest Products

Forest produce comprise of an important part of local economies as they can provide livelihood and resource. Timber is also treated before it is processed as a commodity. Some of the forest resource processing techniques are woodworking, wood preservation and paperboard making.

Woodworking

Artists can use woodworking to create delicate sculptures.

Woodworking is the activity or skill of making items from wood, and includes cabinet making (Cabinetry and Furniture), wood carving, joinery, and carpentry.

History

Along with stone, clay and animal parts, wood was one of the first materials worked by early humans. Microwear analysis of the Mousterian stone tools used by the Neanderthals show that many were used to work wood. The development of civilization was closely tied to the development of increasingly greater degrees of skill in working these materials.

Ancient Egyptian woodworking

Among early finds of wooden tools are the worked sticks from Kalambo Falls, Clacton-on-Sea and Lehringen. The spears from Schöningen (Germany) provide some of the first examples of wooden hunting gear. Flint tools were used for carving. Since Neolithic times, carved wooden vessels are known, for example, from the Linear Pottery culture wells at Kückhofen and Eythra.

Woodworking shop in Germany in 1568, the worker in front is using a bow saw, the one in the background is planing.

Examples of Bronze Age wood-carving include tree trunks worked into coffins from northern Germany and Denmark and wooden folding-chairs. The site of Fellbach-Schmieden in Germany has provided fine examples of wooden animal statues from the Iron Age Wooden idols from the La Tène period are known from a sanctuary at the source of the Seine in France.

Ancient Egypt

There is significant evidence of advanced woodworking in Ancient Egypt. Woodworking is depicted in many extant ancient Egyptian drawings, and a considerable amount of ancient Egyptian furniture (such as stools, chairs, tables, beds, chests) has been preserved. Tombs represent a large collection of these artefacts and the inner coffins found in the tombs were also made of wood. The metal used by the Egyptians for woodworking tools was originally copper and eventually, after 2000 BC bronze as ironworking was unknown until much later.

Commonly used woodworking tools included axes, adzes, chisels, pull saws, and bow drills. Mortise and tenon joints are attested from the earliest Predynastic period. These joints were strengthened using pegs, dowels and leather or cord lashings. Animal glue came to be used only in the New Kingdom period. Ancient Egyptians invented the art of veneering and used varnishes for finishing, though the composition of these varnishes is unknown. Although different native acacias were used, as was the wood from the local sycamore and tamarisk trees, deforestation in the Nile valley resulted in the need for the importation of wood, notably cedar, but also Aleppo pine, boxwood and oak, starting from the Second Dynasty.

Ancient Rome

Woodworking was essential to the Romans. It provided, sometimes the only, material for buildings, transportation, tools, and household items. Wood also provided pipes, dye, waterproofing materials, and energy for heat. Although most examples of Roman woodworking have been lost, the literary record preserved much of the contemporary knowledge. Vitruvius dedicates an entire chapter of his De architectura to timber, preserving many details. Pliny, while not a botanist, dedicated six books of his Natural History to trees and woody plants which provides a wealth of information on trees and their uses.

Ancient China

Damascene woodworkers turning wood for mashrabia and hookass, 19th century.

The progenitors of Chinese woodworking are considered to be Lu Ban (魯班) and his wife Lady Yun, from the Spring and Autumn Period. Lu Ban is said to have introduced the plane, chalk-line, and other tools to China. His teachings were supposedly left behind in the book *Lu Ban Jing* (魯班經, "Manuscript of Lu Ban"). Despite this, it is believed that the text was written some 1500 years after his death. This book is filled largely with descriptions of dimensions for use in building various items such as flower pots, tables, altars, etc., and also contains extensive instructions concerning Feng Shui. It mentions almost nothing of the intricate glue-less and nail-less joinery for which Chinese furniture was so famous.

Micronesian of Tobi, Palau, making a paddle for his *wa* with an adze.

Modern Day

With the advances in modern technology and the demands of industry, woodwork as a field has changed. The development of Computer Numeric Controlled (CNC) Machines, for example, has made us able to mass-produce and reproduce products, faster, with less waste, and often more complex in design than ever before. CNC Routers can carve complicated and highly detailed shapes into flat stock, to create signs or art. Rechargeable power tools speed up creation of many projects and require much less body strength than in the past, when boring multiple holes, for example. Skilled fine woodworking, however, remains a craft pursued by many. There remains demand for hand crafted work such as furniture and arts, however with rate and cost of production, the cost for consumers is much higher.

Materials

Historically, woodworkers relied upon the woods native to their region, until transportation and trade innovations made more exotic woods available to the craftsman. Woods are typically sorted into three basic types: hardwoods typified by tight grain and derived from broadleaf trees, softwoods from coniferous trees, and man-made materials such as plywood and MDF.

Typically furniture such as tables and chairs is made using solid stock, and cabinet/fixture makers employ the use of plywood and other man made panel products.

Paperboard

Corrugated fiberboard made from paperboard

Paperboard is a thick paper-based material. While there is no rigid differentiation between paper and paperboard, paperboard is generally thicker (usually over 0.25 mm, 0.010 in, or 10 points) than paper. According to ISO standards, paperboard is a paper with a grammage above 224 g/m2, but there are exceptions. Paperboard can be single- or multi-ply. Paperboard can be easily cut and formed, is lightweight, and because it is strong, is used in packaging. Another end-use would be graphic printing, such as book and magazine covers or postcards. Sometimes it is referred to as cardboard, which is a generic, lay term used to refer to any heavy paper pulp–based board. Paperboard is also used in fine arts for creating sculptures.

Sculpture made of massive paperboard by Herbert Wetterauer

History

In 1817, the first paperboard carton was produced in England. Folding cartons first emerged around the 1860s and were shipped flat to save space, ready to be set up by customers when they were required. 1879 saw the development of mechanical die cutting and creasing of blanks. In 1911 the first kraft sulphate mill was built in Florida and in 1915 the gable top milk carton was patented and in 1935 the first dairy plant was observed using them. Ovenable paperboard was introduced in 1974.

Terminology and Classification

Terminology and classifications of paperboard are not always uniform. Differences occur depending on specific industry, locale, and personal choice. In general, the following are often used:

- Boxboard or cartonboard – paperboard for folding cartons and rigid set-up boxes
    - Folding boxboard (FBB) – a bending grade capable of being scored and bending without fracture
    - Chipboard – a recycled, low quality board
        - White lined chipboard (WLC) – a white, often clay-coated chipboard
    - Kraft board – a strong virgin fiber board often used for beverage carriers. Often clay- coated for printing
    - Laminated board – a lamination of paperboards and other materials, for example liquid packaging board
    - Solid Bleached Board (SBB) or Solid Bleached Sulphate (SBS) – clean white board used for foods etc. Sulphate refers to the kraft process
    - Solid unbleached board (SUB) – board made from unbleached chemical pulp
- Containerboard – a type of paperboard manufactured for the production of corrugated fiberboard
    - Corrugated medium – the inner fluted portion of corrugated fiberboard
    - Linerboard – a strong stiff board for one or both sides of corrugated boxes. It is the flat covering over the corrugating medium.
- Other
    - Binder's board – a paperboard used in bookbinding for making hardcovers.

Production

Fibrous material is turned into pulp and bleached, to create one or more layers of board, which can be optionally coated for a better surface and/or improved visual appearance. Paperboard are produced on paper machines that can handle higher grammages and several plies.

Raw Materials

The above-mentioned fibrous material can either come from fresh (virgin) sources (e.g. wood) or from recycled waste paper. Around 90% of virgin paper is made from wood pulp. Today paperboard packaging in general, and especially products from certified sustainable sources, are receiving new attention, as manufacturers dealing with environmental, health, and regulatory issues look to renewable resources to meet increasing demand. It is now mandatory in many countries for paper-based packaging to be manufactured wholly or partially from recycled material.

Raw materials include:

- Hardwood: Ca. 0.05 inches (1.3 mm) in length e.g. Birch which has short fibres. It is generally more difficult to work with; however, it does provide higher tensile strength, but lower tear and other strength properties. Although its fibres are not as long and strong as those in softwood, they make for a stiffer product defined by some stifness tests. Hardwood fibres fill the sheet better and therefore make a smoother paper that is more opaque and better for printing. Hardwood makes an excellent corrugating medium.
- Softwood: Ca. 0.13 inches (3.3 mm) in length e.g. Pine and spruce which have typically long fibres and make superior paperboard in services where strength is important. Softwood makes excellent linerboard.
- Recycled: Used paper is collected and sorted and usually mixed with virgin fibres in order to make new material. This is necessary as the recycled fibre often loses strength when reused; the added virgin fibres enhance strength. Mixed waste paper is not usually deinked (skipping the deinking stage) for paperboard manufacture and hence the pulp may contain traces of inks, adhesives, and other residues which together give it a grey colour. Products made of recycled board usually have a less predictable composition and poorer functional properties than virgin fibre-based boards. Health risks have been associated with using recycled material in direct food contact. Swiss studies have shown that recycled material can contain significant portions of mineral oil, which may migrate into packed foods. Mineral oil levels of up to 19.4 mg/kg were found in rice packed in recycled board.
- Others: It is also possible to use the fibres of Sugarcane Bagasse, Straw, Hemp, Cotton, Flax, Kenaf, Abaca and other plant products

Pulping

Two principal methods for extracting fibres from their sources are:

- Chemical pulping uses chemical solutions to convert wood into pulp, yielding around 30% less than mechanical pulping; however, pulp made by the kraft process has superior strength
- Thermo mechanical pulp is a two-stage process which results in a very high yield of wood fibres at the expense of strength.

Bleaching

Pulp used in the manufacture of paperboard can be bleached to decrease colour and increase purity. Virgin fibre pulp is naturally brown in colour, because of the presence of lignin. Recycled paperboard may contain traces of inks, bonding agents and other residue which colors it grey. Although bleaching is not necessary for all end-uses, it is vital for many graphical and packaging purposes. There are various methods of bleaching, which are used according to a number of factors for example, the degree of colour change required, chemicals chosen and method of treatment. There are three categories of bleaching methods:

- Bleaching by delignification using chlorine gas, which is a method that has been largely replaced by procedures which are gentler to the environment such as the use of oxygen as a replacement for the chlorine gas.
- Bleaching by oxidation using chemicals such as chlorine dioxide, hydrogen peroxide or sodium hypochlorite.
- Bleaching by reduction using chemicals such as sodium bisulphite.

Plies

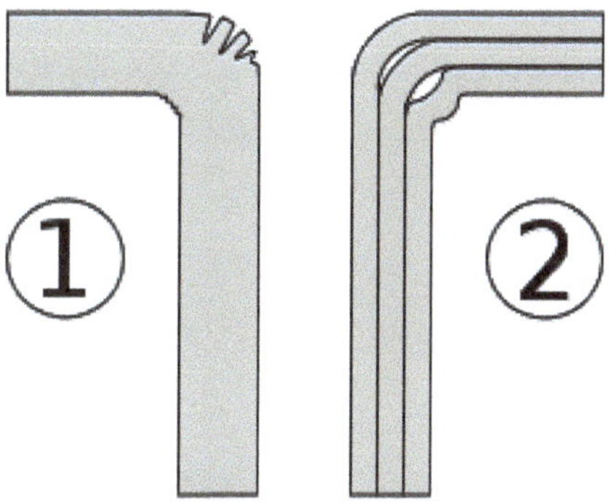

(1) single-ply vs. (2) multi-ply

Multi-ply paperboard generally has higher creasing and folding performance than single-ply as a result of layering different types of pulp into a single product. In cases where the same kind of pulp is being used in several layers, each separate layer is treated and shaped individually in order to create the highest possible quality.

Coating

In order to improve whiteness, smoothness and gloss of paperboard, one or more layers of coating is applied. Coated paper is usually made up of:

- a pigment, which could be china clay, calcium carbonate or titanium dioxide, usually a combination of two
- an adhesive or binder as styrene-butadiene emulsions or starches and water.

Additional components could be OBA (optical brightening agents).

Grades

The DIN Standard 19303 "Paperboard - Terms and grades" (Publication date : 2005-09) defines different grades of paperboard based on the surface treatment (first letter), the main furnish (second letter) and the colour (non-D grade) or bulk (D grade only) (numbering).

First letter (surface treatment)	Second letter (main furnish)	Number
• A = cast-coated • G = pigment coated • U = uncoated	• Z = bleached virgin chemical pulp • C = virgin mechanical pulp • N = unbleached virgin chemical pulp • T = recycled/secondary fibre with white, cream or brown reverse • D = recycled/secondary fibre with grey back	All except D grades: 1. white reverse side 2. cream reverse side 3. brown reverse side D grades only: 1. bulk ≥ 1.45 cm^2/g 2. 1.3 cm^2/g < bulk < 1.45 cm^2/g 3. bulk ≤ 1.3 cm^2/g

Example: GC1 would be a "pigment coated", "virgin mechanical pulp" board with a "white reverse side". Often the used paperboard type would be FBB, which was coated on both sides.

Common Terms

Basis Weight (US): Is the weight in 1,000 square feet (93 m^2) of paperboard.

Brightness: Brightness is a technical term that is defined as the amount of blue-white light that a paper reflects. This property is very subjective and individual to each buyer and end use, as skin colour and food are better reproduced on 'warm' (yellow) whites and not blue whites.

Grammage: The grammage of the paperboard is assessed in accordance ISO 536. Grammage expresses mass per unit area and is measured in g/m^2.

PH: Surface pH is measured on a water extract and is on a scale of 0–14. 0 is acidic, 7 is neutral and 14 is alkaline.

Stiffness: Stiffness is one of the most important properties of paperboard as it affects the ability of cartons to run smoothly through the machine that erects, fills and closes them. Stiffness also gives strength and reduces the propensity of a carton to bulge under the weight of settling flowable contents such as cereals.

Although most paper strength properties increase with increasing sheet density, stiffness does not. A rule of thumb is that stiffness is proportional to the 1.6 power of sheet caliper.

The species of fiber used has an effect on stiffness, other things being equal. Northern softwood species impart superior stiffness compared to southern softwoods.

Other factors which affect board stiffness include coatings and moisture content.

Smoothness: Smoothness is particularly important when being used for printing, the smoother the paperboard, the better the image quality, because of better ink coverage. Smoothness is measured using air leak methods – the greater the rate of air leakage, at a specific air pressure, from under a cylindrical knife placed on the surface, the rougher the surface.

Caliper/Thickness: In the United States caliper is usually expressed in thousandths of an inch (0.001") or points, where a sheet of paperboard with a thickness of 0.024" would be 24 points. In Europe it is often sold in g/m², however the thickness of the board is measured in micron (μm).

Paperboard also tends to be referred to with thickness rather than weight.

Whiteness: It refers ideally to the equal presence of all colours, because a truly white sheet will reflect all wavelengths of visible light equally.

Paperboard Industry

The paperboard sector is mainly looked at in conjunction with the paper industry. The Paper & Paperboard market size (2007) had a value of 630.9 billion USD and a volume of 320.3 million metric tons . 40.1% of that market is European. About 50% of all produced paper is used for packaging, followed by printing and writing. According to ProCarton, the consumption of paper and paperboard seem to correlate with economic trends (GDP). Sales of carton in Europe sum up to around 8 billion Euros worth. Over 1,100 printers produce 5.4 million tonnes of cartonboard yearly. Cartons make up one third of paper and board packaging and 15% of all packaging. A bit more than half (54%) of the European carton is produced using recovered fibre or waste paper.

The paper and paperboard industry is quite energy and capital intensive. Just a coated board machine itself can cost around 90 - 120 million Euros (about 125 - 166 million USD in 11/2011). Economies of scale apply, because of which a few large players often dominate the market place. E.g. in North America the top 5 producers have a market share of 85%.

Fiberboard

Fiberboard (American spelling) or fibreboard (Commonwealth spelling) is a type of engineered wood product that is made out of wood fibers. Types of fiberboard (in order of increasing density) include particle board, medium-density fiberboard (MDF), and hardboard.

Fiberboard is sometimes used as a synonym for particle board, but particle board usually refers to low-density fiberboard. Plywood is not a type of fiberboard, as it is made of thin sheets of wood, not wood fibers or particles. Fiberboard, particularly medium-density fiberboard, is heavily used in the furniture industry. For pieces that will be visible, a veneer of wood is often glued onto fiberboard to give it the appearance of conventional wood.

In the packaging industry, the term "fiberboard" is often used to describe a tough kraft-based paperboard or corrugated fiberboard for boxes.

"Fiberboard" is also an intermediate product, an output of a pulp mill used as input for a paper mill.

Manufacture

Urea-formaldehyde (UF) resins are dominantly used in the medium density fiberboard (MDF) industry because of their low cost and fast curing characteristics. However, pressures on the use of UF resins are mounting steadily due to potential problems associated with formaldehyde emission. On the other hand, phenol-formaldehyde (PF) resins are more durable and do not emit formaldehyde after cure. The industry has traditionally shied away from using PF resins due primarily to their higher cost and much slower curing rate than UF resins. However, the press times for PF-bonded fiberboard can be substantially reduced by manipulating the fiber mat temperatures, molecular weight distribution of PF resins and pressing parameters. As a result, the press times for PF-bonded fiberboard can be made comparable to those for UF-bonded fiberboard. Also, the resin content required for PF-bonded fiberboard is less than 5% to achieve a good board quickly. This is considerably lower than that required for UF-bonded fiberboard.

Certain types of fiberboard can be considered "green" building products. Consisting of bio-based, secondary raw materials (wood chip or sugarcane fibers) recovered from within 100 miles (160 km) of manufacturing facilities, the binding agent used in this type of fiberboard is an all-natural product, consisting of vegetable starch containing no added formaldehydes.

Use

Fiberboard, classified by ASTM C208, Standard Specification for Cellulosic Fiber Insulating Board, has many benefits and is used in residential and commercial construction.

Applications include:

- sound proofing/deadening,
- structural sheathing,
- low-slope roofing,
- sound deadening flooring underlayment,

Fiberboard is also used in the automotive industry to create free-form shapes such as dashboards, rear parcel shelves, and inner door shells. These pieces are usually covered with a skin, foil, or fabric such as cloth, suede, leather, or polyvinyl chloride.

RSI Direct, A bi-weekly e-newsletter covering the roofing, siding and insulation industries, promotes the use of fiberboard as a coverboard in roof systems:

High density coated wood fiber is an ideal cover board, and the industry apparently agrees. More than two billion square feet of this product have already been installed in the U.S. roofing market. In terms of cost and availability, wood fiber is hard to beat.

Wood Veneer

In woodworking, veneer refers to thin slices of wood, usually thinner than 3 mm (1/8 inch), that typically are glued onto core panels (typically, wood, particle board or medium-density fiberboard) to produce flat panels such as doors, tops and panels for cabinets, parquet floors and parts of furniture. They are also used in marquetry. Plywood consists of three or more layers of veneer, each glued with its grain at right angles to adjacent layers for strength. Veneer beading is a thin layer of decorative edging placed around objects, such as jewelry boxes. Veneer is also used to replace decorative papers in Wood Veneer HPL. Veneer is also a type of manufactured board.

Wood veneer RCA Dimensia console TV set

Veneer is obtained either by "peeling" the trunk of a tree or by slicing large rectangular blocks of wood known as flitches. The appearance of the grain and figure in wood comes from slicing through the growth rings of a tree and depends upon the angle at which the wood is sliced. There are three main types of veneer-making equipment used commercially:

- A rotary lathe in which the wood is turned against a very sharp blade and peeled off in one continuous or semi-continuous roll. Rotary-cut veneer is mainly used for plywood, as the appearance is not desirable because the veneer is cut concentric to the growth rings.
- A slicing machine in which the flitch or piece of log is raised and lowered against the blade and slices of the log are made. This yields veneer that looks like sawn pieces of wood, cut across the growth rings; such veneer is referred to as "crown cut".
- A half-round lathe in which the log or piece of log can be turned and moved in such a way as to expose the most interesting parts of the grain, creating a more textured feel and appearance; such veneer is commonly referred to as "rift cut."

Each slicing processes gives a very distinctive type of grain, depending upon the tree species. In any of the veneer-slicing methods, when the veneer is sliced, a distortion of the grain occurs. As it hits the wood, the knife blade creates a "loose" side where the cells have been opened up by the blade, and a "tight" side.

Historically veneers were also sawn, but this is more wasteful of wood. Veneering is an ancient art, dating back to at least the ancient Egyptians who used expensive and rare wood veneers over cheaper timbers to produce their furniture and sarcophagi. During the Roman Empire, Romans also used veneered work in mass quantities.

Producing Wood Veneers

The finest and rarest logs are sent to companies that produce veneer. The advantage to this practice is twofold. First, it provides the most financial gain to the owner of the log. Secondly, and of more importance to the woodworker, it greatly expands the amount of usable wood. While a log used for solid lumber is cut into thick pieces, usually no thinner than 1 1/8 inches (3 cm), veneers are cut as thin as 1/40 of an inch (0.6mm). Depending on the cutting process used by the veneer manufacturer, very little wood is wasted by the saw blade thickness, known as the saw kerf. Accordingly, the yield of a rare grain pattern or wood type is greatly increased, in turn placing less stress on the resource. Some manufacturers even use a very wide knife to "slice off" the thin veneer pieces. In this way, none of the wood is wasted. The slices of veneer are always kept in the order in which they are cut from the log and are often sold this way.

Types of Veneers

There are a few types of veneers available, each serving a particular purpose.

- Raw veneer has no backing on it and can be used with either side facing up. It is important to note that the two sides will appear different when a finish has been applied, due to the cell structure of the wood.
- Paper backed veneer is as the name suggests, veneers that are backed with paper. The advantage to this is it is available in large sizes, or sheets, as smaller pieces are joined together prior to adding the backing. This is helpful for users that do not wish to join smaller pieces of raw veneers together. This is also helpful when veneering curves and columns as the veneer is less likely to crack.
- Phenolic backed veneer is less common and is used for composite, or manmade wood veneers. Due to concern for the natural resource, this is becoming more popular. It too has the advantage of being available in sheets, and is also less likely to crack when being used on curves.
- Laid up veneer is raw veneer that has been joined together to make larger pieces. The process is time-consuming and requires great care, but is not difficult and requires no expensive tools or machinery. Veneers can be ordered through some companies already laid up to any size, shape or design.
- Reconstituted veneer is made from fast-growing tropical species. Raw veneer is cut from a log, and dyed if necessary. Once dyed, the sheets are laminated together to form a block. The block is then sliced so that the edges of the laminated veneer become the "grain" of the reconstituted veneer.
- Wood on Wood Also called 2-ply is a decorative wood veneer face with a utility grade wood backer applied at an opposing direction to the face veneer.

Advantages of Using Veneers

Compared to wood, one of the primary advantages of using veneer is stability. While solid wood can be prone to warping and splitting, because veneer is made of thin layers of wood glued together, the chances of splitting or cracking are reduced. Further, the glue used provides additional strength, making the end result stronger than natural wood.

Some projects built using wood veneer would not be possible to construct using solid lumber, owing to expansion and contraction caused by fluctuation of temperature and humidity. Another advantage of veneer is sustainability—furniture made with wood veneer uses less wood than the same piece of furniture made with solid wood. Further, veneer may also be more readily available than solid wood as exotic hardwood lumber can be scarce and very expensive.

Ecological Characteristics

- Recyclability and renewability: Wood has the least impact on total energy use, greenhouse gasses, air and water pollution, solid waste and ecological resource use. Seventy percent less energy is required in the use of wood compared with any other building materials.
- Sustainability: Using veneer extends the use of a piece of timber. The wood that might be used in one solid piece a few visible cm wide can cover a far greater area when used as a veneer.
- Toxicity: Nontoxic; veneer stores carbon and also maximises the use of harvested wood.

Buying Veneers

Wood veneers are typically sold by the square foot. With the ability to join veneers, even small pieces are usable, resulting in very little waste. Many sources sell small packets of veneers that are sequence matched and are ideal for small projects. These make experimenting and practicing much more economical. It is also possible to buy plywood and other substrates with veneered faces for larger projects consisting of casework.

Wood Preservation

All measures that are taken to ensure a long life of wood fall under the definition wood preservation (timber treatment).

Detail of sample in photo above

In moist and oxygenated soil, there are few treatments that enable vulnerable wood (softwood here) to resist for long against bacterial or fungal degradation

Apart from structural wood preservation measures, there are a number of different (chemical) preservatives and processes (also known as timber treatment, lumber treatment or pressure treatment) that can extend the life of wood, timber, wood structures or engineered wood. These generally increase the durability and resistance from being destroyed by insects or fungus.

History

As proposed by Richardson, treatment of wood has been practised for almost as long as the use of wood itself. There are records of wood preservation reaching back to ancient Greece during Alexander the Great's rule, where bridge wood was soaked in olive oil. The Romans protected their ship hulls by brushing the wood with tar. During the Industrial Revolution wood preservation became a cornerstone of the wood processing industry. Inventors and scientists such as Bethell, Boucherie, Burnett and Kyan made historic developments in wood preservation, with the preservative solutions and processes. Commercial pressure treatment began in the latter half of the 19th century with the protection of railroad crossties using creosote. Treated wood was used primarily for industrial, agricultural, and utility applications, where it is still used, until its use grew considerably (at least in the United States) in the 1970s as homeowners began building decks and backyard projects. Innovation in treated timber products continues to this day, with consumers becoming more interested in less toxic materials.

A modern wharf piling bored by bivalves known as *shipworms*.

Hazards

Wood that has been industrially pressure-treated with approved preservative products poses a limited risk to the public, and should be disposed of properly. On December 31, 2003, the U.S. wood treatment industry stopped treating residential lumber with arsenic and chromium (chromated copper arsenate, or CCA). This was a voluntary agreement with the United States Environmental Protection Agency. CCA was replaced by copper-based pesticides, with exceptions for certain industrial uses. CCA may still be used for outdoor products like utility trailer beds and non-residential construction like piers, docks, and agricultural buildings. Industrial wood preservation chemicals are generally not available directly to the public and may require special approval to import or purchase depending on the product and the jurisdiction where being used. In most countries, industrial wood preservation operations are notifiable industrial activities that require licensing from relevant regulatory authorities such as EPA or equivalent. Reporting and licensing conditions vary widely depending on the particular chemicals used and the country of use.

Although pesticides are used to treat lumber, preserving lumber protects natural resources (in the short term) by enabling wood products to last longer. Previous poor practices in industry have left legacies of contaminated ground and water around wood treatment sites in some cases. However, under currently approved industry practices and regulatory controls such as implemented in Europe, North America, Australia, New Zealand, Japan and elsewhere, environmental impact of these operations should be minimal.

Wood treated with modern preservatives is generally safe to handle given appropriate handling precautions and personal protection measures. However, treated wood may present certain hazards in some circumstances such as during combustion or where loose wood dust particles or other fine toxic residues are generated or where treated wood comes into direct contact with food and agriculture.

Preservatives containing copper in the form of microscopic particles have recently been introduced to the market, usually with "micronized" or "micro" trade names and designations such as MCQ or MCA. While the manufacturers represent that these products are safe and EPA has registered these products, some groups have expressed concerns regarding exposure to engineered sub-micron and nano-sized copper particles. These concerns have led to a debate among advocacy groups and governments on whether special regulation of nanotechnology is warranted. (See nanotechnology and nanotoxicology for additional information.)

The American Wood Protection Association (AWPA), recommends that all treated wood be accompanied by a Consumer Information Sheet (CIS), to communicate safe handling and disposal instructions as well as potential health and environmental hazards of treated wood. Many producers have opted instead to provide Material Safety Data Sheets (MSDS) instead. Although the practice of distributing MSDS instead of CIS is widespread there is an ongoing debate regarding the practice and how best to best communicate potential hazards and hazard mitigation to the end-user. Neither MSDS or the newly adopted International Safety Data Sheets (SDS) are required for treated lumber under current U.S. Federal law.

Businesses and Organizations

American Wood Protection Association

Founded in 1904, the American Wood Protection Association (AWPA), formerly American Wood-Preservers' Association, is a non-profit organization which is the standard setting body for wood preservation standards (including ANSI). AWPA Standards are developed by its technical committees in an open, consensus-based process that involves individuals from all facets of wood preservation: Producers of preservatives and preservative components; producers of treated and untreated wood products; end users of treated wood; engineers, architects and building code officials; government entities, academia, and other groups with a general interest in wood preservation. AWPA's Standards are universally specified for wood preservation in the US, and are recognized worldwide.

AWPA standards help ensure that treated wood products perform satisfactorily for their intended use. They are recognized and used by most, if not all, specifiers of treated wood including electrical utility, marine, road and building construction as well as by local, state and federal governments. "AWPA", "American Wood Protection Association", identifiers of AWPA Standards (e.g., U1, T1, M4, etc.), and Use Category designations (e.g., UC1, UC3B, UC4A, etc.) are AWPA trademarks and the intellectual property of AWPA and its Technical Committees.

Wood preservative systems produced under the AWPA standards system for the residential market are required to be inspected under the stringent American Lumber Standards Committee (ALSC) third party inspection system in order to assure compliance with AWPA standards.

While many wood preservative systems are produced under the AWPA standards system, there are wood preservative products in the market that have not earned AWPA standard status and are not subject to the ALSC inspection system. Compliance with AWPA and ASLC will be noted by the AWPA logo on the product end tags.

Marketing of Wood Preservatives

In general, marketers of wood preservation systems favor certain terminology. For example, the term 'preservative' is used in preference to words such as: chemical, pesticide, fungicide or biocide. And with newer preservatives, the term 'micronized' is favored over the term nanoparticle or nanotechnology, which may raise public safety concerns.

Chemical Preservatives

Chemical preservatives can be classified into three broad categories: water-borne preservatives, oil-borne preservatives, and light organic solvent preservatives (LOSPs). These are discussed in more detail below.

Micronized Copper Technology

Particulate (micronised or dispersed) copper preservative technology has recently been introduced in the US and Europe. In these systems, the copper is ground to micro sized particles and suspended in water rather than being dissolved in a chemical reaction as is the case with other copper products such as ACQ and Copper Azole. There are currently two particulate copper systems in production. One system uses a quat biocide system (known as MCQ) and is a take-off of ACQ. The other uses an azole biocide (known as MCA or μCA-C) and is a take-off of Copper Azole.

Proponents of the particulate copper systems make the case that the particulate copper system perform as well or better than the dissolved copper systems as a wood preservative, but other industry researchers disagree. None of the particulate copper systems have been submitted to the American Wood Protection Association (AWPA) for evaluation, thus the particulate systems should not be used in applications where AWPA standards are required. However, all of the particulate copper systems have been tested and approved for building code requirements by the International Code Council (ICC). The particulate copper systems provide a lighter color than dissolved copper systems such as ACQ or copper azole.

Proponents of the micronized copper systems claim that the systems are subject to third party inspection under a quality monitor program. However, the monitoring program is not subject to oversight by the American Lumber Standards Committee (ALSC) as is required for the AWPA standard systems.

Two particulate copper systems, one marketed as MicroPro and the other as Wolmanized using μCA-C formulation, have achieved Environmentally Preferable Product (EPP) certification. The *EPP* certification was issued by Scientific Certifications Systems (SCS), and is based on a comparative life-cycle impact assessments with an industry standard.

The copper particle size used in the "micronized" copper beads ranges from 1 to 700 nm with an average under 300 nm. Larger particles (such as actual micron-scale particles) of copper do not adequately penetrate the wood cell walls. These micronized preservatives use nano particles of copper oxide or copper carbonate, for which there are alleged safety concerns. An environmental group has recently petitioned EPA to revoke the registration of the micronized copper products citing safety issues.

Alkaline Copper Quaternary

Alkaline copper quaternary (ACQ) is a preservative made of copper, a fungicide, and a quaternary ammonium compound (quat) like didecyl dimethyl ammonium chloride, an insecticide which also augments the fungicidal treatment. ACQ has come into wide use in the US, Europe, Japan and Australia following restrictions on CCA. Its use is governed by national and international standards, which determine the volume of preservative uptake required for a specific timber end use.

Since it contains high levels of copper, ACQ-treated timber is five times more corrosive to common steel. It is necessary to use double-galvanized or stainless steel fasteners in ACQ timber. Use of fasteners meeting or exceeding requirements for ASTM A 153 Class D meet the added requirements for fastener durability. The U.S. began mandating the use of non-arsenic containing wood preservatives for virtually all residential use timber in 2004.

The American Wood Protection Association (AWPA) standards for ACQ require a retention of 0.15 lb/ft^3 (PCF) for above ground use and 0.40 lb/ft^3 for ground contact.

Chemical Specialties, Inc (CSI, now Viance) received U.S. Environmental Protection Agency's Presidential Green Chemistry Challenge Award in 2002 for commercial introduction of ACQ. Its widespread use has eliminated major quantities of arsenic and chromium previously contained in CCA.

Copper Azole

Copper azole preservative (denoted as CA-B and CA-C under American Wood Protection Association/AWPA standards) is a major copper based wood preservative that has come into wide use in Canada, the US, Europe, Japan and Australia following restrictions on CCA. Its use is governed by national and international standards, which determine the volume of preservative uptake required for a specific timber end use.

Copper azole is similar to ACQ with the difference being that the dissolved copper preservative is augmented by an azole co-biocide like Tebuconazole instead of the quat biocide used in ACQ. The azole co-biocide yields a copper azole product that is effective at lower retentions than required for equivalent ACQ performance.

It is marketed widely under the *Wolmanized* brand in North America, and the *Tanalith* brand across Europe and other international markets.

The AWPA standard retention for CA-B is 0.10 lb/ft^3 for above ground applications and 0.21 lb/ft^3 for ground contact applications. Type C copper azole, denoted as CA-C, has been introduced under the Wolmanized brand. The AWPA standard retention for CA-C is 0.06 lb/ft^3 for above ground applications and 0.15 lb/ft^3 for ground contact applications.

The copper azole preservative incorporates organic triazoles such as tebuconazole or propiconazole as the co-biocide, which are also used to protect food crops. The general appearance of wood treated with copper azole preservative is similar to CCA with a green colouration.

Copper Naphthenate

Copper Naphthenate a Danish invention of 1911, has been used as a preservative and water repellant for more than 100 years. Copper Naphthenate has been used effectively for many applications including: fence posts, canvas, nets, greenhouses, utility poles, railroad ties, beehives, and wooden structures in ground contact. Copper Naphthenate is registered with the EPA as a non-restricted use pesticide, so there is no federal applicators licensing requirements for its use as a wood preservative. Copper Naphthenate can be applied by brush, dip, or pressure treatment.

The University of Hawaii has found that Copper Naphthenate in wood at loadings of 1.5 lbs per cubic foot is resistant to Formosan termite attack. On February 19, 1981 the Federal Register outlined the EPA's position regarding the health risks associated the various wood preservatives. As a result, the National Park Service recommended the use of copper naphthenate in its facilities as an approved substitute for pentachlorophenal, creosote, and inorganic arsenicals. A 50-year study presented to AWPA in 2005 by Mike Freeman and Douglas Crawford says, "This study reassessed the condition of the treated wood posts in southern Mississippi, and statistically calculated the new expected post life span. It was determined that commercial wood preservatives, like pentachlorophenol in oil, creosote, and copper naphthenate in oil, provided excellent protection for posts, with life spans now calculated to exceed 60 years. Surprisingly, creosote and penta treated posts at 75% of the recommended AWPA retention, and Copper Naphthenate at 50% of the required AWPA retention, gave excellent performance in this AWPA Hazard Zone 5 site. Untreated southern pine posts lasted 2 years in this test site."

The AWPA M4 Standard for the care of preservative-treated wood products, reads, "The appropriateness of the preservation system for field treatment shall be determined by the type of preservative originally used to protect the product and the availability of a field treatment preservative. Because many preservative products are not packaged and labeled for use by the general public, a system different from the original treatment may need to be utilized for field treatment. Users shall carefully read and follow the instructions and precautions listed on the product label when using these materials. Copper Naphthenate preservatives containing a minimum of 2.0% copper metal are recommended for material originally treated with copper naphthenate, pentachlorophenol, creosote, creosote solution or waterborne preservatives." The M4 Standard has been adopted by the International Code Council's (ICC) International Building Code (IBC) section 2303.1.8 Preservative-treated Wood, and International Residential Code (IRC) R319.1.1 Field Treatment. The American Association of State Highway and Transportation Officials AASHTO has also adopted the AWPA M4 Standard.

A waterborne copper naphthenate is sold to consumers under the tradename QNAP 5W. Oilborne copper napthenates with 1% copper as metal solutions are sold to consumers under the tradenames Copper Green, and Wolmanized Copper Coat, a 2% copper as metal solution is sold under the tradename Tenino.

Chromated Copper Arsenate (CCA)

In CCA treatment, copper is the primary fungicide, arsenic is a secondary fungicide and an insecticide, and chromium is a fixative which also provides ultraviolet (UV) light resistance. Recognized for the greenish tint it imparts to timber, CCA is a preservative that was extremely common for many decades.

In the pressure treatment process, an aqueous solution of CCA is applied using a vacuum and pressure cycle, and the treated wood is then stacked to dry. During the process, the mixture of oxides reacts to form insoluble compounds, helping with leaching problems.

The process can apply varying amounts of preservative at varying levels of pressure to protect the wood against increasing levels of attack. Increasing protection can be applied (in increasing order of attack and treatment) for: exposure to the atmosphere, implantation within soil, or insertion into a marine environment.

In the last decade concerns were raised that the chemicals may leach from the wood into surrounding soil, resulting in concentrations higher than naturally occurring background levels. A study cited in *Forest Products Journal* found 12–13% of the chromated copper arsenate leached from treated wood buried in compost during a 12-month period. Once these chemicals have leached from the wood, they are likely to bind to soil particles, especially in soils with clay or soils that are more alkaline than neutral. In the United States the US Consumer Product Safety Commission issued a report in 2002 stating that exposure to arsenic from direct human contact with CCA treated wood may be higher than was previously thought. On 1 January 2004, the Environmental Protection Agency (EPA) in a voluntary agreement with industry began restricting the use of CCA in treated timber in residential and commercial construction, with the exception of shakes and shingles, permanent wood foundations, and certain commercial applications. This was in an effort to reduce the use of arsenic and improve environmental safety, although the EPA were careful to point out that they had not concluded that CCA treated wood structures in service posed an unacceptable risk to the community. The EPA did not call for the removal or dismantling of existing CCA treated wood structures.

In Australia, the Australian Pesticides and Veterinary Medicines Authority (APVMA) restricted the use of CCA preservative for treatment of timber used in certain applications from March 2006. CCA may no longer be used to treat wood used in 'intimate human contact' applications such as children's play equipment, furniture, residential decking and handrailing. Use for low contact residential, commercial and industrial applications remains unrestricted, as does its use in all other situations. The APVMA decision to restrict the use of CCA in Australia was a precautionary measure, even though the report found no evidence that demonstrated CCA treated timber posed unreasonable risks to humans in normal use. Similarly to the US EPA, the APVMA did not recommend dismantling or removal of existing CCA treated wood structures.

In Europe, Directive 2003/2/EC restricts the marketing and use of arsenic, including CCA wood treatment. CCA treated wood is not permitted to be used in residential or domestic constructions. It is permitted for use in various industrial and public works, such as bridges, highway safety fencing, electric power transmission and telecommunications poles. In the United Kingdom waste timber treated with CCA was classified in July 2012 as hazardous waste by the Department for the Environment, Food and Rural Affairs.

Other Copper Compounds

These include copper HDO (Bis-(N-cyclohexyldiazeniumdioxy)-copper or CuHDO), copper chromate, copper citrate, acid copper chromate, and ammoniacal copper zinc arsenate (ACZA). The CuHDO treatment is an alternative to CCA, ACQ and CA used in Europe and in approval stages for United States and Canada. ACZA is generally used for marine applications.

Borate Preservatives

Boric acid, oxides and salts (borates) are effective wood preservatives and are supplied under numerous brand names throughout the world. One of the most common compounds used is disodium octaborate tetrahydrate (commonly abbreviated DOT). Borate treated wood is of low toxicity to humans, and does not contain copper or other heavy metals. However, unlike most other preservatives, borate compounds do not become fixed in the wood and can be partially leached out if exposed repeatedly to water that flows away rather than evaporating (evaporation leaves the borate behind so is not a problem). Even though leaching will not normally reduce boron concentrations below effective levels for preventing fungal growth, borates should not be used where they will be exposed to repeated rain, water or ground contact unless the exposed surfaces are treated to repel water. Zinc-borate compounds are less suspectible to leaching than sodium-borate compounds, but are still not recommended for below-ground use unless the timber is first sealed. Recent interest in low toxicity timber for residential use, along with new regulations restricting some wood preservation agents, has resulted in a resurgence of the use of borate treated wood for floor beams and internal structural members. Researchers at CSIRO in Australia have developed organoborates which are much more resistant to leaching, while still providing timber with good protection from termite and fungal attack. The cost of the production of these modified borates will limit their widespread take-up but they are likely to be suitable for certain niche applications, especially where low mammalian toxicity is of paramount importance.

PTI Wood Preservatives

Recent concerns about the health and environmental impacts of metallic wood preservatives have created a market interest in non-metallic wood preservatives such as Propiconazole-Tebuconazole-Imidicloprid better known as PTI. The American Wood Protection Association (AWPA) standards for PTI require a retention of 0.018 lb/ft3 (PCF) for above ground use and 0.013 lb/ft3 when applied in combination with a wax stabilizer. The AWPA has not developed a standard for a PTI ground contact preservative, so PTI is currently limited to above ground applications such as decks. All three of the PTI components are also used in food crop applications. The very low required retentions for PTI pressure treated wood further limits impacts plus substantinally decreases the freight costs and associated environmental impacts for shipping preservative components to the pressure treating plants.

The PTI preservative imparts very little color to the wood. Producers generally add a color agent or a trace amount of copper solution so as to identify the wood as pressure treated and to better match the color of other pressure treated wood products. The PTI wood products are very well adapted for paint and stain applications with no bleed-through. The addition of the wax stabilizer allows a lower preservative retention plus substantially reduces the tendency of wood to warp and split as it dries. In combination with normal deck maintenance and sealer applications, the stabilizer helps maintain appearance and performance over time. PTI pressure treated wood products are no more corrosive than untreated wood and are approved for all types of metal contact, including aluminum.

PTI pressure treated wood products are relatively new to the market place and are not yet widely available in building supply stores. However, there are some suppliers selling PTI products for delivery anywhere in the US on a job lot order basis.

Sodium Silicate-based Preservatives

Sodium silicate is produced by fusing sodium carbonate with sand or heating both ingredients under pressure. It has been in use since the 19th century. It can be a deterrent against insect attack and possesses minor flame-resistant properties; however, it is easily washed out of wood by moisture, forming a flake-like layer on top of the wood.

Timber Treatment Technology, LLC, markets TimberSIL®, a sodium silicate wood preservative. The TimberSIL® proprietary process surrounds the wood fibers with a protective, non-toxic, amorphous glass matrix. The result is a product the company calls "Glass Wood," a material that is Class A fire-retardant, chemically inert, rot and decay resistant, and superior in strength to untreated wood.Another option is to use plywood, fire resistant plywood

Potassium Silicate-based Preservatives

There are a number of European natural paint fabricants that have developed potassium silicate (potassium waterglass) based preservatives. They frequently include boron compounds, cellulose, lignin and other plant extracts. They are a surface application with a minimal impregnation for internal use.

Bifenthrin Spray Preservatives

In Australia, a water-based bifenthrin preservative has been developed to improve the insect resistance of timber. As this preservative is applied by spray, it only penetrates the outer 2 mm of the timber cross-section. Concerns have been raised as to whether this thin-envelope system will provide protection against insects in the longer term, particularly when exposed to sunlight for extended periods.

Fire Retardant Treated

This treated wood utilizes a fire retardant chemical that remains stable in high temperature environments. The fire retardant is applied under pressure at a wood treating plant like the preservatives described above, or applied as a surface coating.

In both cases, treatment provides a physical barrier to flame spread. The treated wood chars but does not oxidize. Effectively this creates a convective layer that transfers flame heat to the wood in a uniform way which significantly slows the progress of fire to the material. There are several commercially available wood-based construction materials using pressure-treatment (such as those marketed in the United States and elsewhere under the trade names of "FirePro", "Burnblock" 'Woodsafe, Dricon', 'D-Blaze,' and 'Pyro-Guard'), as well as factory-applied coatings under the trade names of 'PinkWood' and 'BluWood'. Some site-applied coatings as well as brominated fire retardants have lost favor due to safety concerns as well as concerns surrounding the consistency of application. Specialized treatments also exist for wood used in weather-exposed applications.

The only impregnation-applied fire retardant commercially available in Australia (marketed under the tradename 'Guardian') used calcium formate as a 'powerful wood modifying agent', but was removed from sale in early 2010 for unspecified reasons.

Oil-borne Preservatives

These include pentachlorophenol ("penta") and creosote. They emit a strong petrochemical odor and are generally not used in consumer products.

Coal-tar Creosote

Creosote was the first wood preservative to gain industrial importance more than 150 years ago and it is still widely used today for protection of industrial timber components where long service life is essential. Creosote is a tar-based preservative that is commonly used for utility poles and railroad ties (UK: railway sleepers). Creosote is one of the oldest wood preservatives, and was originally derived from a wood distillate, but now, virtually all creosote is manufactured from the distillation of coal tar. Creosote is regulated as a pesticide, and is not usually sold to the general public.

Linseed Oil

In recent years in Australia and New Zealand, linseed oil has been incorporated in preservative formulations as a solvent and water repellent to "envelope treat" timber. This involves just treating the outer 5 mm of the cross-section of a timber member with preservative (e.g., permethrin 25:75), leaving the core untreated. While not as effective as CCA or LOSP methods, envelope treatments are significantly cheaper, as they use far less preservative. Major preservative manufacturers add a blue (or red) dye to envelope treatments. Blue colored timber is for use south of the Tropic of Capricorn and red for elsewhere. The colored dye also indicates that the timber is treated for resistance to termites/white ants. There is an ongoing promotional campaign in Australia for this type of treatment.

Other Emulsions

Light Organic Solvent Preservatives (LOSP)

This class of timber treatments use white spirit, or light oils such as kerosene, as the solvent carrier to deliver preservative compounds into timber. Synthetic pyrethroids are typically used as an insecticide, such as permethrin, bifenthrin or deltamethrin. In Australia and New Zealand, the most common formulations use Permethrin as an insecticide, and Propaconazole and Tebuconazole as fungicides. While still using a chemical preservative, this formulation contains no heavy-metal compounds.

With the introduction of strict volatile organic compound (VOC) laws in the European Union, LOSPs have disadvantages due to the high cost and long process times associated with vapour-recovery systems. LOSPs have been emulsified into water-based solvents. While this does significantly reduce VOC emissions, the timber swells during treatment, removing many of the advantages of LOSP formulations.

Epoxy

Various Epoxy resins usually thinned with a solvent like acetone or MEK can be used to both preserve and seal wood.

New Technologies

Wood Acetylation

This bridge made from acetylated wood near Sneek, the Netherlands, is designed to carry heavy traffic.

Chemical modification of wood at the molecular level has been used to improve its performance properties. Many chemical reaction systems for the modification of wood, especially those using various types of anhydrides, have been published; however, the reaction of wood with acetic anhydride has been the most studied.

The physical properties of any material are determined by its chemical structure. Wood contains an abundance of chemical groups called *free hydroxyls*. Free hydroxyl groups readily absorb and release water according to changes in the climatic conditions to which they are exposed. This is the main reason why wood's dimensional stability is impacted by swelling and shrinking. It is also believed that the digestion of wood by enzymes initiates at the free hydroxyl sites, which is one of the principal reasons why wood is prone to decay.

Acetylation effectively changes the free hydroxyls within wood into acetyl groups. This is done by reacting the wood with acetic anhydride, which comes from acetic acid. When free hydroxyl groups are transformed to acetyl groups, the ability of the wood to absorb water is greatly reduced, rendering the wood more dimensionally stable and, because it is no longer digestible, extremely durable. In general, softwoods naturally have an acetyl content from 0.5 to 1.5% and more durable hardwoods from 2 to 4.5%. Acetylation takes wood well beyond these levels with corresponding benefits. These include an extended coatings life due to acetylated wood acting as a more stable substrate for paints and translucent coatings. Acetylated wood is non-toxic and does not have the environmental issues associated with traditional preservation techniques.

The acetylation of wood was first done in Germany in 1928 by Fuchs. In 1946, Tarkow, Stamm and Erickson first described the use of wood acetylation to stabilize wood from swelling in water. Since the 1940s, many laboratories around the world have looked at acetylation of many different types of woods and agricultural resources.

In spite of the vast amount of research on chemical modification of wood, and, more specifically, on the acetylation of wood, commercialization did not come easily. The first patent on the acetylation of wood was filed by Suida in Austria in 1930. Later, in 1947, Stamm and Tarkow filed a patent on the acetylation of wood and boards using pyridine as a catalyst. In 1961, the Koppers Company published a technical bulletin on the acetylation of wood using no catalysis, but with an organic cosolvent In 1977, in Russia, Otlesnov and Nikitina came close to commercialization, but the process was discontinued, presumably because cost-effectiveness could not be achieved. In 2007, Titan Wood, a London-based company, with production facilities in The Netherlands, achieved cost-effective commercialization and began large-scale production of acetylated wood under the trade name "Accoya".

Natural Preservatives

Copper Plating

Copper plating or Copper sheathing is the practice of covering wood most usually wooden hulls of ships with copper metal. As metallic copper is both repellent and toxic to fungus, insects such as termites, and marine bi-valves this would preserve the wood and also act as an anti-fouling measure to prevent aquatic life from attaching to the ship's hull and reducing a ship's speed and maneuverability.

Naturally Rot-resistant Woods

These species are resistant to decay in their natural state, due to high levels of organic chemicals called *extractives*, mainly polyphenols. Extractives are chemicals that are deposited in the heartwood of certain tree species as they convert sapwood to heartwood. Huon pine (*Lagarostrobos franklinii*), merbau (*Intsia bijuga*), ironbark (*Eucalyptus* spp.), tōtara (*Podocarpus totara*), puriri (*Vitex lucens*), kauri (*Agathis australis*), and many cypresses, such as coast redwood (*Sequoia sempervirens*) and western red cedar (*Thuja plicata*), fall in this category. However, many of these species tend to be prohibitively expensive for general construction applications.

Huon pine was used for ship hulls in the 19th century, but over-harvesting and Huon pine's extremely slow growth rate makes this now a speciality timber. Huon pine is so rot resistant, that fallen trees from many years ago are still commercially valuable.

Merbau is still a popular decking timber and has a long life in above ground applications, but it is logged in an unsustainable manner and is too hard and brittle for general use.

Ironbark is a good choice where available. It is harvested from both old-growth and plantation in Australia and is highly resistant to rot and termites. It is most commonly used for fence posts and house stumps.

Eastern red cedar (*Juniperus virginiana*) and black locust (*Robinia pseudoacacia*) have long been used for rot-resistant fence posts and rails in eastern United States, with the black locust also planted in modern times in Europe. Coast redwood is commonly used for similar applications in the western United States.

Tōtara and puriri were used extensively in New Zealand during the European colonial era when native forests were "mined", even as fence posts of which many are still operating. Tōtara was used by the Māori to build large *waka* (canoes). Today, they are specialty timbers as a result of their scarcity, although lower grade stocks are sold for landscaping use.

Kauri is a superb timber for building the hulls and decks of boats. It too is now a specialty timber and ancient logs (in excess of 3,000 years) that have been mined from swamps are used by wood turners and furniture makers.

The natural durability or rot and insect resistance of wood species is always based on the heartwood (or "truewood"). The sapwood of all timber species should be considered to be non-durable without preservative treatment.

Tung Oil

Tung oil has been used for hundreds of years in China, where it was used as a preservative for wood ships. The oil penetrates the wood, and then hardens to form an impermeable hydrophobic layer up to 5 mm into the wood. As a preservative it is effective for exterior work above and below ground, but the thin layer makes it less useful in practice. It is not available as a pressure treatment.

Heat Treatments

By going beyond kiln drying wood, heat treatment may make timber more durable. By heating timber to a certain temperature, it may be possible to make the wood fibre less appetizing to insects.

Heat treatment can also improve the properties of the wood with respect to water, with lower equilibrium moisture, less moisture deformation, and weather resistance. It is weather-resistant enough to be used unprotected, in facades or in kitchen tables, where wetting is expected.

There are four similar heat treatments — Westwood, developed in the United States; Retiwood, developed in France; Thermowood, developed in Finland by VTT; and Platowood, developed in The Netherlands. These processes autoclave the treated wood, subjecting it to pressure and heat, along with nitrogen or water vapour to control drying in a staged treatment process ranging from 24 to 48 hours at temperatures of 180 °C to 230 °C depending on timber species. These processes increase the durability, dimensional stability and hardness of the treated wood by at least one class; however, the treated wood is darkened in colour, and there are changes in certain mechanical characteristics: Specifically, the modulus of elasticity is increased to 10%, and the modulus of rupture is diminished by 5% to 20%; thus, the treated wood requires drilling for nailing to avoid splitting the wood. Certain of these processes cause less of an impact than others in their the mechanical effects upon the treated wood. Wood treated with this process is often used for cladding or siding, flooring, furniture and windows.

For the control of pests that may be harbored in wood packaging material (i.e. crates and pallets), the ISPM 15 requires heat treatment of wood to 56 °C for 30 minutes to receive the HT stamp. This is typically required to ensure the killing of the pine wilt nematode and other kinds of wood pests that could be transported internationally.

Mud Treatment

Wood and bamboo can be buried in mud to help protect them from insects and decay. This practice is used widely in Vietnam to build farm houses consisting of a wooden structural frame, a bamboo roof frame and bamboo with mud mixed with rice hay for the walls. While wood in contact with soil will generally decompose more quickly than wood not in contact with it, it is possible that the predominantly clay soils prevalent in Vietnam provide a degree of mechanical protection against insect attack, which compensates for the accelerated rate of decay.

Also, since wood is only subject to bacterial decay under specific temperature and moisture content ranges, submerging it in water-saturated mud can retard decay, by saturating the wood's internal cells beyond their moisture decay range.

Application Processes

Introduction and History

Probably the first attempts made to protect wood from decay and insect attack consisted of brushing or rubbing preservatives onto the surfaces of the treated wood. Through trial and error the most effective preservatives and application processes were slowly determined. In the Industrial Revolution, demands for such things as telegraph poles and railroad ties (UK: railway sleepers) helped to fuel an explosion of new techniques that emerged in the early 19th century. The sharpest rise in inventions took place between 1830 and 1840, when Bethell, Boucherie, Burnett and Kyan were making wood-preserving history. Since then, numerous processes have been introduced or existing processes improved. The goal of modern-day wood preservation is to ensure a deep, uniform penetration with reasonable cost, without endangering the environment. The most widespread application processes today are those using artificial pressure through which many woods are being effectively treated, but several species (such as spruce, Douglas-fir, larch, hemlock and fir) are very resistant to impregnation. With the use of incising, the treatment of these woods has been somewhat successful but with a higher cost and not always satisfactory results. One can divide the wood-preserving methods roughly into either non-pressure processes or pressure processes.

Non-pressure Processes

There are numerous non-pressure processes of treating wood which vary primarily in their procedure. The most common of these treatments involve the application of the preservative by means of brushing or spraying, dipping, soaking, steeping or by means of hot and cold bath. There is also a variety of additional methods involving charring, applying preservatives in bored holes, diffusion processes and sap displacement.

Brush and Spray Treatments

Brushing preservatives is a long-practised method and often used in today's carpentry workshops. Technological developments mean it is also possible to spray preservative over the surface of the timber. Some of the liquid is drawn into the wood as the result of capillary action before the spray runs off or evaporates, but unless puddling occurs penetration is limited and may not be suitable for long-term weathering. By using the spray method, coal-tar creosote, oil-borne solutions and water-borne salts (to some extent) can also be applied. A thorough brush or spray treatment with coal-tar creosote can add 1 to 3 years to the lifespan of poles or posts. Two or more coats provide better protection than one, but the successive coats should not be applied until the prior coat has dried or soaked into the wood. The wood should be seasoned before treatment.

Dipping

Dipping consists of simply immersing the wood in a bath of creosote or other preservative for a few seconds or minutes. Similar penetrations to that of brushing and spraying processes are achieved. It has the advantage of minimizing hand labor. It requires more equipment and larger quantities of preservative and is not adequate for treating small lots of timber. Usually the dipping process is useful in the treatment of window sashes and doors. Treatment with copper salt preservatives is no longer allowed with this method.

Steeping

In this process the wood is submerged in a tank of water-preservative mix, and allowed to soak for a longer period of time (several days to weeks). This process was developed in the 19th century by John Kyan. The depth and retention achieved depends on factors such as species, wood moisture, preservative and soak duration. The majority of the absorption takes place during the first two or three days, but will continue at a slower pace for an indefinite period. As a result, the longer the wood can be left in the solution, the better treatment it will receive. When treating seasoned timber, both the water and the preservative salt soak into the wood, making it necessary to season the wood a second time. Posts and poles can be treated directly on endangered areas, but should be treated at least 30 cm (0.98 ft) above the future ground level.

The depth obtained during regular steeping periods varies from 5 to 10 mm (0.20 to 0.39 in) up to 30 mm (1.2 in) by sap pine. Due to the low absorption, solution strength should be somewhat stronger than that in pressure processes, around 5% for seasoned timber and 10% for green timber (because the concentration slowly decreases as the chemicals diffuse into the wood). The solution strength should be controlled continually and, if necessary, be corrected with the salt additive. After the timber is removed from the treatment tank, the chemical will continue to spread within the wood if it has sufficient moisture content. The wood should be weighed down and piled so that the solution can reach all surfaces. (Sawed materials stickers should be placed between every board layer.) This process finds minimal use despite its former popularity in continental Europe and Great Britain.

Kyanizing

Named after John Howard Kyan, who patented this process in England in 1833, Kyanizing consists of steeping wood in a 0.67% mercuric chloride preservative solution.

Gedrian's Bath

Patented by Charles A. Seely, this process achieves treatment by immersing seasoned wood in successive baths of hot and cold preservatives. During the hot baths, the air expands in the timbers. When the timbers are changed to the cold bath (the preservative can also be changed) a partial vacuum is created within the lumen of the cells, causing the preservative to be drawn into the wood. Some penetration occurs during the hot baths, but most of it takes place during the cold baths. This cycle is repeated with a significant time reduction compared to other steeping processes. Each bath may last 4 to 8 hours or in some cases longer. The temperature of the preservative in the hot bath should be between 60 to 110 °C (140 to 230 °F) and 30 to 40 °C (86 to 104 °F) in the cold bath (depending on preservative and tree species). The average penetration depths achieved with this process ranges from 30 to 50 mm (1.2 to 2.0 in). Both preservative oils and water-soluble salts can be used with this treatment. Due to the longer treatment periods, this method finds little use in the commercial wood preservation industry today.

Preservative Precipitation

As explained in Uhlig's Corrosion Handbook, this process involves two or more chemical baths that undergo a reaction with the cells of the wood, and result in the precipitation of preservative into the wood cells. Two chemicals commonly employed in this process are Copper Ethanolamine, and Sodium Dimethyldithiocarbamate, which reacts to precipitate Copper Dimetyldithiocarbamate. The precipitated preservative is very resistant to leeching. Since its use in the mid 1990s, it has been discontinued in the United States of America, however it never saw commercialization in Canada.

Pressure Processes

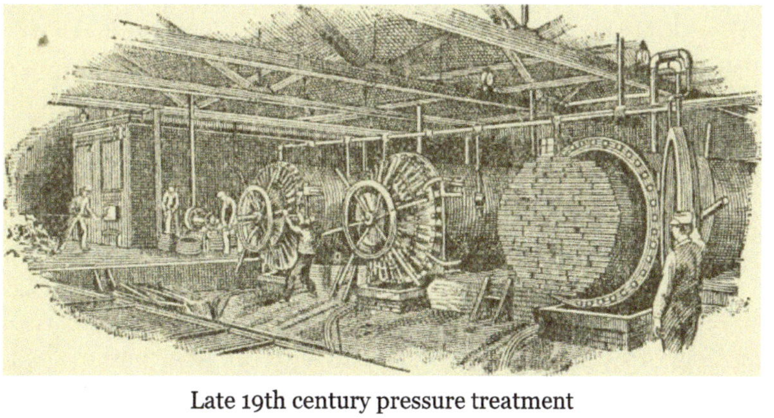

Late 19th century pressure treatment

Pressure processes are the most permanent method around today in preserving timber life. Pressure processes are those in which the treatment is carried out in closed cylinders with applied pressure and/or vacuum. These processes have a number of advantages over the non-pressure methods. In most cases, a deeper and more uniform penetration and a higher absorption of preservative is achieved. Another advantage is that the treating conditions can be controlled so that retention and penetration can be varied. These pressure processes can be adapted to large-scale production. The high initial costs for equipment and the energy costs are the biggest disadvantages. These treatment methods are used to protect ties, poles and structural timbers and find use throughout the world today. The various pressure processes that are used today differ in details, but the general method is in all cases the same. The treatment is carried out in cylinders. The timbers are loaded onto special tram cars, so called *buggies* or *bogies*, and into the cylinder. These cylinders are then set under pressure often with the addition of higher temperature. As final treatment, a vacuum is frequently used to extract excess preservatives. These cycles can be repeated to achieve better penetration.

LOSP treatments often use a vacuum impregnation process. This is possible because of the lower viscosity of the white-spirit carrier used.

Full-cell Process

In the full-cell process, the intent is to keep as much of the liquid absorbed into the wood during the pressure period as possible, thus leaving the maximum concentration of preservatives in the treated area. Usually, water solutions of preservative salts are employed with this process, but it is also possible to impregnate wood with oil. The desired retention is achieved by changing the strength of the solution. William Burnett patented this development in 1838 of full-cell impregnation with water solutions. The patent covered the use of zinc chloride on water basis, also known as *Burnettizing*. A full-cell process with oil was patented in 1838 by John Bethell. His patent described the injection of tar and oils into wood by applying pressure in closed cylinders. This process is still used today with some improvements.

Fluctuation Pressure Process

Contrary to the static full-cell and empty-cell processes, the fluctuation process is a dynamic process. By this process the pressure inside the impregnation cylinder changes between pressure and vacuum within a few seconds. There have been inconsistent claims that through this process it is possible to reverse the pit closure by spruce. However, the best results that have been achieved with this process by spruce do not exceed a penetration deeper than 10 mm (0.39 in). Specialized equipment is necessary and therefore higher investment costs are incurred.

Boucherie Process

Developed by Dr. Boucherie of France in 1838, this approach consisted of attaching a bag or container of preservative solution to a standing or a freshly cut tree with bark, branches, and leaves still attached, thereby injecting the liquid into the sap stream. Through transpiration of moisture from the leaves the preservative is drawn upward through the sapwood of the tree trunk.

The modified Boucherie process consists of placing freshly cut, unpeeled timbers onto declining skids, with the stump slightly elevated, then fastening watertight covering caps or boring a number of holes into the ends, and inserting a solution of copper sulfate or other waterborne preservative into the caps or holes from an elevated container. Preservative oils tend to not penetrate satisfactorily by this method. The hydrostatic pressure of the liquid forces the preservative lengthwise into and through the sapwood, thus pushing the sap out of the other end of the timber. After a few days, the sapwood is completely impregnated; unfortunately little or no penetration takes place in the heartwood. Only green wood can be treated in this manner. This process has found considerable usage to impregnate poles and also larger trees in Europe and North America, and has experienced a revival of usage to impregnate bamboo in countries such as Costa Rica, Bangladesh, India and the state of Hawaii.

High-pressure Sap Displacement System

Developed in the Philippines, this method (abbreviated HPSD) consists of a cylinder pressure cap made from a 3 mm thick mild steel plate secured with 8 sets of bolts, a 2-HP diesel engine, and a pressure regulator with 1.4–14 kg/m^2 capacity. The cap is placed over the stump of a pole, tree or bamboo and the preservative is forced into the wood with pressure from the engine.

Incising

First tested and patented by Kolossvary, Haltenberger, and Berdenich of Austria in 1911 and 1912 (U.S. pats. 1,012,207 and 1,018,624) with several improvements from O. P. M. Goss, D. W. Edwards and J. H. Mansfield among others, this process consists of making shallow, slit-like holes in the surfaces of material to be treated, so that deeper and more uniform penetration of preventative may be obtained. The term *incising* or perforating comes from the Latin *incidere*, a compound of *in* and *caedere* (to cut). Incisions made in sawed material usually are parallel with the grain of the wood. This process is common in North America (since the 1950s), where Douglas-fir products and pole butts of various species are prepared before treatment. It is most useful for woods that are resistant to side penetration, but allow preservative transport along the grain. In the region in which it is produced, it is common practice to incise all sawed Douglas-fir 3 in (76 mm) or more in thickness before treatment.

Unfortunately, the impregnation of spruce, the most important structural timber in large areas in Europe, has shown that unsatisfactory treatment depths have been achieved with impregnation. The maximum penetration of 2 mm (0.079 in) is not sufficient to protect wood in weathered positions. The present-day incising machines consist essentially of four revolving drums fitted with teeth or needles or with lasers that burn the incisions into the wood. Preservatives can be spread along the grain up to 20 mm (0.79 in) in radial and up to 2 mm (0.079 in) in tangential and radial direction.

In North America, where smaller timber dimensions are common, incision depths of 4 to 6 mm (0.16 to 0.24 in) have become standard. In Europe, where larger dimensions are widespread, incision depths of 10 to 12 mm (0.39 to 0.47 in) are necessary. The incisions are visible and often considered to be wood error. Incisions by laser are significantly smaller than those of spokes or needles. The costs for each process type are approximately for spoke/conventional all-round incising €0.50/m^2, by laser incising €3.60/m^2 and by needle incision €1.00/m^2. (Figures originate from the year 1998 and may vary from present day prices.)

Microwaving

An alternative increases the permeability of timber using microwave technology. There is some concern that this method may adversely affect the structural performance of the material. Research in this area has been conducted by the Cooperative Research Centre at the University of Melbourne, Australia.

Charring

Charring of timber results in surfaces which are fire-resistant, insect-resistant and proof against weathering. Wood surfaces are ignited using a hand-held burner or moved slowly across a fire. The charred surface is then cleaned using a steel brush to remove loose bits and to expose the grain. Oil or varnish may be applied if required.

Timber Recycling

Demolishers pulling timber from an old wool store in Sydney, which will later be re-used for timber flooring

Timber recycling or wood recycling is the process of turning waste timber into usable products. Recycling timber is a practice that was popularized in the early 1990s as issues such as deforestation and climate change prompted both timber suppliers and consumers to turn to a more sustainable timber source. Recycling timber is the environmentally friendliest form of timber production and is very common in countries such as Australia and New Zealand where supplies of old wooden structures are plentiful. Timber can be chipped down into wood chips which can be used to power homes or power plants.

Benefits

Recycling timber has become popular due to its image as an environmentally friendly product. Common belief among consumers is that by purchasing recycled wood, the demand for "green timber" will fall and ultimately benefit the environment. Greenpeace also view recycled timber as an environmentally friendly product, citing it as the most preferable timber source on their website. The arrival of recycled timber as a construction product has been important in both raising industry and consumer awareness towards deforestation and promoting timber mills to adopt more environmentally friendly practices.

Example of recycled timber as a final product

Drawbacks

Some hurdles facing the widespread adoption of recycled timber: sometimes the ends of wall studs need to be trimmed off to stop decay and cracking, thus resulting in a shorter piece of wood; this trimming may result in pieces of wood that do not meet building codes. Though the price may be less than for new wood, the process of selecting usable pieces of salvaged wood, pulling out nails, and refinishing for a new use can be laborious and time-consuming. Demolition must happen in such a way as to preserve as much of the timber as possible in a building, which means more time spent dismantling a building rather than just tearing it down quickly. The trade in recycled timber is not well-established everywhere, so a reliable supply of usable wood may be hard to come by for builders. There may be a stigma associated with using "used" or "cheap" wood that is perceived to be of not as high quality as "new" wood. Not all pieces of wood in a dismantled building will fit in a new building, and it may be cheaper and easier, from a design and labor perspective, to simply get new wood (ex: wood from a 6-foot (1.8 m) deck being used in a 7-foot (2.1 m) deck). Of course, none of these issues are insurmountable, and they are issues of convenience and logistics rather than structural integrity, but many builders find it easier and less time-consuming to simply get new wood in standard uniform sizes.

Recycling Timber

Recycled timber most commonly comes from old buildings, bridges and wharfs, where it is carefully stripped out and put aside by demolishers. At the same time any usable dimension stone is set aside for reuse. The demolishers then sell the salvaged timber to merchants who then re-mill the timber by manually scanning it with a metal detector, which allows the timber to be de-nailed and sawn to size. Once re-milled the timber is commonly sold to consumers in the form of timber flooring, beams and decking.

Recycled timber salvaged from a demolished building in Trat, Thailand, is applied to make a new roof.

Examples

Use of recycling timber is not new. As early as 1948, the 100 metres tall tower of Golm transmitter near Potsdam, Germany was built from recycled timber. It stood for 31 years. It was common to reuse wood of dismantled radio towers in the 1930s in Germany, e.g. the former tower of Koblenz radio transmitter was built of wood from a dismantled tower, which carried a T-antenna at Transmitter Muehlacker. The upper parts of the 157-metre-tall wood tower of Ismaning radio transmitter, which stood for 49 years, were built of wood from a smaller radio tower dismantled in 1934.

Non-timber Forest Product

Non-timber forest products (NTFPs), also special, non-wood, minor, alternative and secondary forest products, are useful substances, materials and/or commodities obtained from forests which do not require harvesting (logging) trees. They include game animals, fur-bearers, nuts, seeds, berries, mushrooms, oils, foliage, medicinal plants, peat, fuelwood, and forage.

Mahua seeds

Tendu patta (leaf) collection

Research on NTFPs have focused on their commodifiability for rural incomes and markets, as an expression of traditional knowledge or as a livelihood option for rural household needs, and, as a key component of sustainable forest management and conservation strategies. All research promote forest products as valuable commodities and tools that can promote the conservation of forests.

Definitions

There is a wide variety of NTFPs, including mushrooms, huckleberries, ferns, transplants, seed cones, piñon seeds, tree nuts, moss, maple syrup, cork, cinnamon, rubber, tree oils and resins, and ginseng. The United Kingdom's Forestry Commission defines NTFPs as "any biological resources found in woodlands except timber," and Forest Harvest, part of the Reforesting Scotland project, defines them as "materials supplied by woodlands - except the conventional harvest of timber." These definitions include wild and managed game, fish and insects. NTFPs are commonly grouped into categories such as floral greens, decoratives, medicinal plants, foods, flavors and fragrances, fibers, and saps and resins.

Other terms similar to NTFPs include *special, non-wood, minor, alternative* and *secondary forest products*. NTFPs in particular highlight forest products which are of value to local people and communities but have been overlooked in the wake of forest management priorities (for example, timber production and animal forage). In recent decades, interest has grown in using NTFPs as alternatives or supplements to forest management practices. In some forest types, under the right political and social conditions, forests can be managed to increase NTFP diversity and, consequently, to increase biodiversity and potentially economic diversity. Black truffle cultivation in the mediterranean area is a good example of a high profitability when well managed.

Uses

The harvest of NTFPs remains widespread throughout the world. People from a wide range of socio-economic, geographical and cultural contexts harvest NTFPs for a number of purposes, including but not limited to: household subsistence, maintenance of cultural and familial traditions, spiritual fulfillment as well as physical and emotional well-being, scientific learning and income. Other terms synonymous with harvesting include *wild-crafting, gathering, collecting* and *foraging*. NTFPs also serve as raw materials for industries ranging from large-scale floral greens suppliers and pharmaceutical companies to micro-enterprises centred upon a wide variety of activities (such as basket-making, woodcarving and the harvest and processing of various medicinal plants).

Economic Importance

It is difficult to estimate the contribution of NTFPs to national or regional economies as there is a lack of broad-based systems for tracking the combined value of the hundreds of products that make up various NTFP industries. One exception to this is the maple syrup industry, which in 2002 in the US alone yielded 1.4 million US gallons (5,300 m^3) worth USD$38.3 million. In temperate forests such as in the US, wild edible mushrooms such as matsutake, medicinal plants such as ginseng, and floral greens such as salal and *sword fern* are multimillion-dollar industries. While these high-value species may attract the most attention, a diversity of NTFPs can be found in most forests of the world.

In tropical forests, for example, NTFPs can be an important source of income that can supplement farming and/or other activities. A value-analysis of the Amazon rainforest in Peru found that exploitation of NTFPs could yield higher net revenue per hectare than would timber harvest of the same area, while still conserving vital ecological services. Their economic, cultural and ecological value, when considered in aggregate, makes managing NTFPs an important component of sustainable forest management and the conservation of biological and cultural diversity.

Research

Research on NTFPs have focused on three perspectives: NTFPs as a commodity with a focus on rural incomes and markets, as an expression of traditional knowledge or as a livelihood option for rural household needs, and, finally, as a key component of sustainable forest management and conservation strategies. These perspectives promote forest products as valuable commodities and important tools that can promote the conservation of forests. In some contexts, the gathering and use of NTFPs can be a mechanism for poverty alleviation and local development.

Bamboo

The bamboos are a subfamily (Bambusoideae) of flowering perennial evergreen plants in the grass family Poaceae.

Giant bamboos are the largest members of the grass family. In bamboo, the internodal regions of the stem are usually hollow and the vascular bundles in the cross section are scattered throughout the stem instead of in a cylindrical arrangement. The dicotyledonous woody xylem is also absent. The absence of secondary growth wood causes the stems of monocots, including the palms and large bamboos, to be columnar rather than tapering.

Bamboos include some of the fastest-growing plants in the world, due to a unique rhizome-dependent system. Certain species of bamboo can grow 91 cm (3 ft) within a 24-hour period, at a rate of almost 4 cm (1.5 in) an hour (a growth around 1 mm every 90 seconds, or one inch every 40 minutes). Bamboos are of notable economic and cultural significance in South Asia, Southeast Asia and East Asia, being used for building materials, as a food source, and as a versatile raw product. Bamboo has a higher compressive strength than wood, brick, or concrete and a tensile strength that rivals steel.

The word bamboo comes from the Kannada term *bambu*, which was introduced to English through Indonesian and Malay.

Systematics and Taxonomy

Bamboos have long been considered the most primitive grasses, mostly because of the presence of bracteate, indeterminate inflorescences, "pseudospikelets", and flowers with three lodicules, six stamens, and three stigmata. Following more recent molecular phylogenetic research, many tribes and genera of grasses formerly included in the Bambusoideae are now classified in other subfamilies, e.g. the Anomochlooideae, the Puelioideae, and the Ehrhartoideae. The subfamily in its current sense belongs to the BOP clade of grasses, where it is sister to the Pooideae (bluegrasses and relatives).

The bamboos contains three clades classified as tribes, and these strongly correspond with geographic divisions representing the New World herbaceous species (Olyreae), tropical woody bamboos (Bambuseae), and temperate woody bamboos (Arundinarieae). The woody bamboos do not form a monophyletic group; instead, the tropical woody and herbaceous bamboos are sister to the temperate woody bamboos. Altogether, more than 1,400 species are placed in 115 genera.

Tribe Olyreae (Herbaceous Bamboos)

21 genera:

- Subtribe Buergersiochloineae

  one genus: Buergersiochloa

- Subtribe Olyrineae

  20 genera: *Agnesia, Arberella, Cryptochloa, Diandrolyra, Ekmanochloa, Froesiochloa, Lithachne, Maclurolyra, Mniochloa, Olyra (plant), Pariana, Parianella, Parodiolyra, Piresia, Piresiella, Raddia, Raddiella, Rehia, Reitzia, Sucrea.*

Tribe Bambuseae (Tropical Woody Bamboos)

91 genera:

- Subtribe Arthrostylidiinae

  13 genera: *Actinocladum, Alvimia, Apoclada, Arthrostylidium, Athroostachys, Atractantha, Aulonemia (Matudacalamus), Colanthelia, Elytrostachys, Glaziophyton, Merostachys, Myriocladus, Rhipidocladum.*

- Subtribe Bambusinae

  10 genera: *Bambusa (Dendrocalamopsis), Bonia (Monocladus), Dendrocalamus (Klemachloa, Oreobambos, Oxynanthera* or *Sinocalamus), Dinochloa, Gigantochloa, Holttumochloa, Kinabaluchloa (Maclurochloa, Soejatmia), Melocalamus, Sphaerobambos, Thyrsostachys.*

- Subtribe Chusqueinae

  2 genera: *Chusquea, Neurolepis.*

- Subtribe Guaduinae

  4 genera: *Eremocaulon, Guadua, Olmeca, Otatea.*

- Subtribe Melocanninae

  9 genera: *Cephalostachyum, Davidsea, Leptocanna, Melocanna, Neohouzeaua, Ochlandra, Pseudostachyum, Schizostachyum, Teinostachyum.*

- Subtribe Nastinae

  6 genera: *Decaryochloa, Greslania, Hickelia, Hitchcockella, Nastus, Perrierbambus.*

- Subtribe Racemobambodinae

  one genus: *Racemobambos (Neomicrocalamus, Vietnamosasa)*

- Subtribe Shibataeinae

  8 genera: *Chimonobambusa, Indosasa, Phyllostachys, Qiongzhuea, Semiarundinaria (Brachystachyum), Shibataea, Sinobambusa, Temburongia* (incertae sedis).

Tribe Arundinarieae (Temperate Woody Bamboos)

16 genera: *Acidosasa, Ampelocalamus, Arundinaria, Borinda, Chimonocalamus, Drepanostachyum (Himalayacalamus), Fargesia, Ferrocalamus, Gaoligongshania, Gelidocalamus, Indocalamus, Oligostachyum, Pseudosasa, Sasa, Thamnocalamus, Yushania.*

Distribution

Worldwide distribution of bamboos

Bamboo species are found in diverse climates, from cold mountains to hot tropical regions. They occur across East Asia, from 50°N latitude in Sakhalin through to Northern Australia, and west to India and the Himalayas. They also occur in sub-Saharan Africa, and in the Americas from the USA mid-Atlantic states south to Argentina and Chile, reaching their southernmost point at 47°S latitude. Continental Europe is not known to have any native species of bamboo.

Recently, some attempts have been made to grow bamboo on a commercial basis in the Great Lakes region of east-central Africa, especially in Rwanda. In the United States, several companies are growing, harvesting, and distributing species such as *Phyllostachys nigra* (Henon) and *Phyllostachys edulis* (Moso).

Ecology

Bamboo forest in Taiwan

Bamboos include some of the fastest-growing plants on Earth, with reported growth rates up to 91 cm (36 in) in 24 hours. However, the growth rate is dependent on local soil and climatic conditions, as well as species, and a more typical growth rate for many commonly cultivated bamboos in temperate climates is in the range of 3–10 cm (1.2–3.9 in) per day during the growing period. Primarily growing in regions of warmer climates during the late Cretaceous period, vast fields existed in what is now Asia. Some of the largest timber bamboo can grow over 30 m (98 ft) tall, and be as large as 25–30 cm (9.8–11.8 in) in diameter. However, the size range for mature bamboo is species-dependent, with the smallest bamboos reaching only several inches high at maturity. A typical height range that would cover many of the common bamboos grown in the United States is 4.6–12 m (15–39 ft), depending on species. Anji County of China, known as the "Town of Bamboo", provides the optimal climate and soil conditions to grow, harvest, and process some of the most valued bamboo poles available worldwide.

Bamboo forest in Arashiyama

Bamboo forest in Kwa-Zulu Natal

Unlike all trees, individual bamboo culms emerge from the ground at their full diameter and grow to their full height in a single growing season of three to four months. During these several months, each new shoot grows vertically into a culm with no branching out until the majority of the mature height is reached. Then, the branches extend from the nodes and leafing out occurs. In the next year, the pulpy wall of each culm slowly hardens. During the third year, the culm hardens further. The shoot is now considered a fully mature culm. Over the next 2–5 years (depending on species), fungus begins to form on the outside of the culm, which eventually penetrates and overcomes the culm. Around 5–8 years later (species- and climate-dependent), the fungal growths cause the culm to collapse and decay. This brief life means culms are ready for harvest and suitable for use in construction within about three to seven years. Individual bamboo culms do not get any taller or larger in diameter in subsequent years than they do in their first year, and they do not replace any growth lost from pruning or natural breakage. Bamboo has a wide range of hardiness depending on species and locale. Small or young specimens of an individual species produce small culms initially. As the clump and its rhizome system mature, taller and larger culms are produced each year until the plant approaches its particular species limits of height and diameter.

Many tropical bamboo species die at or near freezing temperatures, while some of the hardier temperate bamboos can survive temperatures as low as −29 °C (−20 °F). Some of the hardiest bamboo species can be grown in places as cold as USDA plant hardiness zones 5–6, although they typically defoliate and may even lose all above-ground growth, yet the rhizomes survive and send up shoots again the next spring. In milder climates, such as USDA zone 8 and above, some hardy bamboo may remain fully leafed out year-round.

Bamboo Canopy

Mass Flowering

Flowering bamboo

Most bamboo species flower infrequently. In fact, many only flower at intervals as long as 65 or 120 years. These taxa exhibit mass flowering (or gregarious flowering), with all plants in a particular 'cohort' flowering over a several-year period. Any plant derived through clonal propagation from this cohort will also flower regardless of whether it has been planted in a different location. The longest mass flowering interval known is 130 years, and it is for the species *Phyllostachys bambusoides* (Sieb. & Zucc.). In this species, all plants of the same stock flower at the same time, regardless of differences in geographic locations or climatic conditions, and then the bamboo dies. The lack of environmental impact on the time of flowering indicates the presence of some sort of "alarm clock" in each cell of the plant which signals the diversion of all energy to flower production and the cessation of vegetative growth. This mechanism, as well as the evolutionary cause behind it, is still largely a mystery.

One hypothesis to explain the evolution of this semelparous mass flowering is the predator satiation hypothesis, which argues that by fruiting at the same time, a population increases the survival rate of its seeds by flooding the area with fruit, so even if predators eat their fill, seeds will still be left over. By having a flowering cycle longer than the lifespan of the rodent predators, bamboos can regulate animal populations by causing starvation during the period between flowering events. Thus, the death of the adult clone is due to resource exhaustion, as it would be more effective for parent plants to devote all resources to creating a large seed crop than to hold back energy for their own regeneration.

Phyllostachys glauca 'Yunzhu' in flower

Another, the fire cycle hypothesis, states that periodic flowering followed by death of the adult plants has evolved as a mechanism to create disturbance in the habitat, thus providing the seedlings with a gap in which to grow. This argues that the dead culms create a large fuel load, and also a large target for lightning strikes, increasing the likelihood of wildfire. Because bamboos can be aggressive as early successional plants, the seedlings would be able to outstrip other plants and take over the space left by their parents.

However, both have been disputed for different reasons. The predator satiation hypothesis does not explain why the flowering cycle is 10 times longer than the lifespan of the local rodents, something not predicted. The bamboo fire cycle hypothesis is considered by a few scientists to be unreasonable; they argue that fires only result from humans and there is no natural fire in India. This notion is considered wrong based on distribution of lightning strike data during the dry season throughout India. However, another argument against this is the lack of precedent for any living organism to harness something as unpredictable as lightning strikes to increase its chance of survival as part of natural evolutionary progress.

More recently, a mathematical explanation for the extreme length of the flowering cycles has been offered, involving both the stabilizing selection implied by the predator satiation hypothesis and others, and the fact that plants that flower at longer intervals tend to release more seeds. The hypothesis claims that bamboo flowering intervals grew by integer multiplication. A mutant bamboo plant flowering at a noninteger multiple of its population's flowering interval would release its seeds alone, and would not enjoy the benefits of collective flowering (such as protection from predators). However, a mutant bamboo plant flowering at an integer multiple of its population's flowering interval would release its seeds only during collective flowering events, and would release more seeds than the average plant in the population. It could, therefore, take over the population, establishing a flowering interval that is an integer multiple of the previous flowering interval. The hypothesis predicts that observed bamboo flowering intervals should factorize into small prime numbers.

The mass fruiting also has direct economic and ecological consequences, however. The huge increase in available fruit in the forests often causes a boom in rodent populations, leading to increases in disease and famine in nearby human populations. For example, devastating consequences occur when the *Melocanna bambusoides* population flowers and fruits once every 30–35 years around the Bay of Bengal. The death of the bamboo plants following their fruiting means the local people lose their building material, and the large increase in bamboo fruit leads to a rapid increase in rodent populations. As the number of rodents increases, they consume all available food, including grain fields and stored food, sometimes leading to famine. These rats can also carry dangerous diseases, such as typhus, typhoid, and bubonic plague, which can reach epidemic proportions as the rodents increase in number. The relationship between rat populations and bamboo flowering was examined in a 2009 *Nova* documentary "Rat Attack".

In any case, flowering produces masses of seeds, typically suspended from the ends of the branches. These seeds give rise to a new generation of plants that may be identical in appearance to those that preceded the flowering, or they may produce new cultivars with different characteristics, such as the presence or absence of striping or other changes in coloration of the culms.

Several bamboo species are never known to set seed even when sporadically flowering has been reported. *Bambusa vulgaris*, *Bambusa balcooa*, and *Dendrocalamus stocksii* are common examples of such bamboo.

Animal Diet

Soft bamboo shoots, stems, and leaves are the major food source of the giant panda of China, the red panda of Nepal, and the bamboo lemurs of Madagascar. Rats eat the fruits as described above. Mountain gorillas of Africa also feed on bamboo, and have been documented consuming bamboo sap which was fermented and alcoholic; chimpanzees and elephants of the region also eat the stalks.

Bamboo is the main food of the giant panda, making up 99% of its diet.

The larvae of the bamboo borer (the moth *Omphisa fuscidentalis*) of Laos, Myanmar, Thailand, and Yunnan Province, China, feeds off the pulp of live bamboo. In turn, these caterpillars are considered a local delicacy.

Cultivation

Phyllostachys stalk, Bamboo forest in Isère / France

Bamboo foliage with yellow stems (probably *Phyllostachys aurea*)

Bamboo foliage with black stems (probably *Phyllostachys nigra*)

Commercial Timber

Timber is harvested from both cultivated and wild stands, and some of the larger bamboos, particularly species in the genus *Phyllostachys*, are known as "timber bamboos".

Harvesting

Bamboo used for construction purposes must be harvested when the culms reach their greatest strength and when sugar levels in the sap are at their lowest, as high sugar content increases the ease and rate of pest infestation.

Harvesting of bamboo is typically undertaken according to these cycles:

1) Lifecycle of the culm: As each individual culm goes through a 5– to 7-year lifecycle, culms are ideally allowed to reach this level of maturity prior to full capacity harvesting. The clearing out or thinning of culms, particularly older decaying culms, helps to ensure adequate light and resources for new growth. Well-maintained clumps may have a productivity three to four times that of an unharvested wild clump. Consistent with the lifecycle described above, bamboo is harvested from two to three years through to five to seven years, depending on the species.

2) Annual cycle: As all growth of new bamboo occurs during the wet season, disturbing the clump during this phase will potentially damage the upcoming crop. Also during this high-rainfall period, sap levels are at their highest, and then diminish towards the dry season. Picking immediately prior to the wet/growth season may also damage new shoots. Hence, harvesting is best a few months prior to the start of the wet season.

3) Daily cycle: During the height of the day, photosynthesis is at its peak, producing the highest levels of sugar in sap, making this the least ideal time of day to harvest. Many traditional practitioners believe the best time to harvest is at dawn or dusk on a waning moon.

Leaching

Leaching is the removal of sap after harvest. In many areas of the world, the sap levels in harvested bamboo are reduced either through leaching or postharvest photosynthesis. Examples of this practice include:

1. Cut bamboo is raised clear of the ground and leaned against the rest of the clump for one to two weeks until leaves turn yellow to allow full consumption of sugars by the plant.
2. A similar method is undertaken, but with the base of the culm standing in fresh water, either in a large drum or stream to leach out sap.
3. Cut culms are immersed in a running stream and weighted down for three to four weeks.
4. Water is pumped through the freshly cut culms, forcing out the sap (this method is often used in conjunction with the injection of some form of treatment).

In the process of water leaching, the bamboo is dried slowly and evenly in the shade to avoid cracking in the outer skin of the bamboo, thereby reducing opportunities for pest infestation.

Durability of bamboo in construction is directly related to how well it is handled from the moment of planting through harvesting, transportation, storage, design, construction, and maintenance. Bamboo harvested at the correct time of year and then exposed to ground contact or rain will break down just as quickly as incorrectly harvested material.

Ornamental Bamboos

The two general patterns for the growth of bamboo are "clumping" (sympodial) and "running" (monopodial). Clumping bamboo species tend to spread slowly, as the growth pattern of the rhizomes is to simply expand the root mass gradually, similar to ornamental grasses. "Running" bamboos, though, need to be controlled during cultivation because of their potential for aggressive behavior. They spread mainly through their roots and/or rhizomes, which can spread widely underground and send up new culms to break through the surface. Running bamboo species are highly variable in their tendency to spread; this is related to both the species and the soil and climate conditions. Some can send out runners of several metres a year, while others can stay in the same general area for long periods. If neglected, over time, they can cause problems by moving into adjacent areas.

Bamboos seldom and unpredictably flower, and the frequency of flowering varies greatly from species to species. Once flowering takes place, a plant declines and often dies entirely. Although a few species of bamboo are always in flower at any given time, collectors desiring to grow specific bamboo typically obtain their plants as divisions of already-growing plants, rather than waiting for seeds to be produced.

Regular observations indicate major growth directions and locations. Once the rhizomes are cut, they are typically removed; however, rhizomes take a number of months to mature, and an immature, severed rhizome usually ceases growing if left in-ground. If any bamboo shoots come up outside of the bamboo area afterwards, their presence indicates the precise location of the removed rhizome. The fibrous roots that radiate from the rhizomes do not produce more bamboo.

Bamboo growth is also controlled by surrounding the plant or grove with a physical barrier. Typically, concrete and specially rolled HDPE plastic are the materials used to create the barrier, which is placed in a 60– to 90-cm-deep ditch around the planting, and angled out at the top to direct the rhizomes to the surface. (This is only possible if the barrier is installed in a straight line.) If the containment area is small, this method can be detrimental to ornamental bamboo, as the bamboo within can become rootbound and start to display the signs of any unhealthy containerized plant. In addition, rhizomes can escape over the top, or beneath the barrier if it is not deep enough. Strong rhizomes and tools can penetrate plastic barriers, so care must be taken. In small areas, regular maintenance may be the best method for controlling the running bamboos. Barriers and edging are unnecessary for clump-forming bamboos, although these may eventually need to have portions removed if they become too large.

The ornamental plant sold in containers and marketed as "lucky bamboo" is actually an entirely unrelated plant, *Dracaena sanderiana*. It is a resilient member of the lily family that grows in the dark, tropical rainforests of Southeast Asia and Africa. Lucky bamboo has long been associated with the Eastern practice of *feng shui* and images of the plant widely available on the Web are often used to depict bamboo. On a similar note, Japanese knotweed is also sometimes mistaken for a bamboo, but it grows wild and is considered an invasive species. *Phyllostachys* species of bamboo are also considered invasive and illegal to sell or propagate in some areas of the US.

Uses

Culinary

Although the shoots (new culms that come out of the ground) of bamboo contain a toxin taxiphyllin (a cyanogenic glycoside) that produces cyanide in the gut, proper processing renders them edible. They are used in numerous Asian dishes and broths, and are available in supermarkets in various sliced forms, in both fresh and canned versions. The golden bamboo lemur ingests many times the quantity of the taxiphyllin-containing bamboo that would kill a human.

The bamboo shoot in its fermented state forms an important ingredient in cuisines across the Himalayas. In Assam, India, for example, it is called *khorisa*. In Nepal, a delicacy popular across ethnic boundaries consists of bamboo shoots fermented with turmeric and oil, and cooked with potatoes into a dish that usually accompanies rice (*alu tama* (आलु तामा) in Nepali).

Unprocessed bamboo shoots in a Japanese market

Khao lam (Thai: ข้าวหลาม) is glutinous rice with sugar and coconut cream cooked in specially prepared bamboo sections of different diameters and lengths

In Indonesia, they are sliced thin and then boiled with *santan* (thick coconut milk) and spices to make a dish called *gulai rebung*. Other recipes using bamboo shoots are *sayur lodeh* (mixed vegetables in coconut milk) and *lun pia* (sometimes written *lumpia*: fried wrapped bamboo shoots with vegetables). The shoots of some species contain toxins that need to be leached or boiled out before they can be eaten safely.

Pickled bamboo, used as a condiment, may also be made from the pith of the young shoots.

The sap of young stalks tapped during the rainy season may be fermented to make *ulanzi* (a sweet wine) or simply made into a soft drink. Bamboo leaves are also used as wrappers for steamed dumplings which usually contains glutinous rice and other ingredients.

Pickled bamboo shoots (Nepali: *tama*) are cooked with black-eyed beans as a delicacy food in Nepal. Many Nepalese restaurant around the world serve this dish as *aloo bodi tama*. Fresh bamboo shoots are sliced and pickled with mustard seeds and turmeric and kept in glass jar in direct sunlight for the best taste. It is used alongside many dried beans in cooking during winters. Baby shoots (Nepali: *tusa*) of a very different variety of bamboo (Nepali: Nigalo) native to Nepal is cooked as a curry in hilly regions.

In Sambalpur, India, the tender shoots are grated into juliennes and fermented to prepare *kardi*. The name is derived from the Sanskrit word for bamboo shoot, *karira*. This fermented bamboo shoot is used in various culinary preparations, notably *amil*, a sour vegetable soup. It is also made into pancakes using rice flour as a binding agent. The shoots that have turned a little fibrous are fermented, dried, and ground to sand-sized particles to prepare a garnish known as *hendua*. It is also cooked with tender pumpkin leaves to make sag green leaves.

In Konkani cuisine, the tender shoots (*kirlu*) are grated and cooked with crushed jackfruit seeds to prepare *kirla sukke*.

The empty hollow in the stalks of larger bamboo is often used to cook food in many Asian cultures. Soups are boiled and rice is cooked in the hollows of fresh stalks of bamboo directly over a flame. Similarly, steamed tea is sometimes rammed into bamboo hollows to produce compressed forms of Pu-erh tea. Cooking food in bamboo is said to give the food a subtle but distinctive taste.

In addition, bamboo is frequently used for cooking utensils within many cultures, and is used in the manufacture of chopsticks. In modern times, some see bamboo tools as an ecofriendly alternative to other manufactured utensils.

Medicine

Bamboo is used in traditional Chinese medicine for treating infections and healing. In northern Indian state of Assam, the fermented bamboo paste known as *khorisa* is known locally as a folk remedy for the treatment of impotence, infertility, and menstrual pains.

Construction

Bamboo has long been used as an assembly material in Hong Kong because of its versatility.

Bamboo, like true wood, is a natural composite material with a high strength-to-weight ratio useful for structures.

In its natural form, bamboo as a construction material is traditionally associated with the cultures of South Asia, East Asia, and the South Pacific, to some extent in Central and South America, and by extension in the aesthetic of Tiki culture. In China and India, bamboo was used to hold up simple suspension bridges, either by making cables of split bamboo or twisting whole culms of sufficiently pliable bamboo together. One such bridge in the area of Qian-Xian is referenced in writings dating back to 960 AD and may have stood since as far back as the third century BC, due largely to continuous maintenance.

Bamboo has also long been used as scaffolding; the practice has been banned in China for buildings over six stories, but is still in continuous use for skyscrapers in Hong Kong. In the Philippines, the nipa hut is a fairly typical example of the most basic sort of housing where bamboo is used;

the walls are split and woven bamboo, and bamboo slats and poles may be used as its support. In Japanese architecture, bamboo is used primarily as a supplemental and/or decorative element in buildings such as fencing, fountains, grates, and gutters, largely due to the ready abundance of quality timber.

House made entirely of bamboo

Bamboo scaffolding can reach great heights.

Various structural shapes may be made by training the bamboo to assume them as it grows. Squared sections of bamboo are created by compressing the growing stalk within a square form. Arches may similarly be created by forcing the bamboo's growth into the desired form, costing much less than it would to obtain the same shape with regular wood timber. More traditional forming methods, such as the application of heat and pressure, may also be used to curve or flatten the cut stalks.

Bamboo can be cut and laminated into sheets and planks. This process involves cutting stalks into thin strips, planing them flat, and boiling and drying the strips; they are then glued, pressed, and finished. Long used in China and Japan, entrepreneurs started developing and selling laminated bamboo flooring in the West during the mid-1990s; products made from bamboo laminate, including flooring, cabinetry, furniture, and even decorations, are currently surging in popularity, transitioning from the boutique market to mainstream providers such as Home Depot. The bamboo goods industry (which also includes small goods, fabric, etc.) is expected to be worth $25 billion by 2012. The quality of bamboo laminate varies among manufacturers and varies according to the maturity of the plant from which it was harvested (six years being considered the optimum); the sturdiest products fulfill their claims of being up to three times harder than oak hardwood while others may be softer than standard hardwood.

Bamboo intended for use in construction should be treated to resist insects and rot. The most common solution for this purpose is a mixture of borax and boric acid. Another process involves boiling cut bamboo to remove the starches that attract insects.

Bamboo pavilion in the Shenzhen Biennale

Bamboo has been used as reinforcement for concrete in those areas where it is plentiful, though dispute exists over its effectiveness in the various studies done on the subject. Bamboo does have the necessary strength to fulfil this function, but untreated bamboo will swell with water absorbed from the concrete, causing it to crack. Several procedures must be followed to overcome this shortcoming.

Several institutes, businesses, and universities are researching the use of bamboo as an ecological construction material. In the United States and France, it is possible to get houses made entirely of bamboo, which are earthquake- and cyclone-resistant and internationally certified. In Bali, Indonesia, an international K-12 school, the Green School, is constructed entirely of bamboo, for its beauty and advantages as a sustainable resource. Three ISO standards are given for bamboo as a construction material.

In parts of India, bamboo is used for drying clothes indoors, both as a rod high up near the ceiling to hang clothes on, and as a stick wielded with acquired expert skill to hoist, spread, and to take down the clothes when dry. It is also commonly used to make ladders, which apart from their normal function, are also used for carrying bodies in funerals. In Maharashtra, the bamboo groves and forests are called *Veluvana*, the name *velu* for bamboo is most likely from Sanskrit, while *vana* means forest.

Furthermore, bamboo is also used to create flagpoles for saffron-coloured, Hindu religious flags, which can be seen fluttering across India, especially in Bihar and Uttar Pradesh, as well as in Guyana and Suriname in South America.

Bamboo was used for the structural members of the India pavilion at Expo 2010 in Shanghai. The pavilion is the world's largest bamboo dome, about 34 m (112 ft) in diameter, with bamboo beams/members overlaid with a ferro-concrete slab, waterproofing, copper plate, solar PV panels, a small windmill, and live plants. A total of 30 km (19 mi) of bamboo was used. The dome is supported on 18-m-long steel piles and a series of steel ring beams. The bamboo was treated with borax and boric acid as a fire retardant and insecticide and bent in the required shape. The bamboo sections were joined with reinforcement bars and concrete mortar to achieve the necessary lengths.

Textiles

Since the fibers of bamboo are very short (less than 3 mm (0.12 in)), they are not usually transformed into yarn by a natural process. The usual process by which textiles labeled as being made of bamboo are produced uses only rayon made from the fibers with heavy employment of chemicals. To accomplish this, the fibers are broken down with chemicals and extruded through mechanical spinnerets; the chemicals include lye, carbon disulfide, and strong acids. Retailers have sold both end products as "bamboo fabric" to cash in on bamboo's current ecofriendly cachet; however, the Canadian Competition Bureau and the US Federal Trade Commission, as of mid-2009, are cracking down on the practice of labeling bamboo rayon as natural bamboo fabric. Under the guidelines of both agencies, these products must be labeled as rayon with the optional qualifier "from bamboo".

As a Writing Surface

Bamboo was in widespread use in early China as a medium for written documents. The earliest surviving examples of such documents, written in ink on string-bound bundles of bamboo strips (or "slips"), date from the fifth century BC during the Warring States period. However, references in earlier texts surviving on other media make it clear that some precursor of these Warring States period bamboo slips was in use as early as the late Shang period (from about 1250 BC).

Bamboo or wooden strips were the standard writing material during the Han dynasty, and excavated examples have been found in abundance. Subsequently, paper began to displace bamboo and wooden strips from mainstream uses, and by the fourth century AD, bamboo slips had been largely abandoned as a medium for writing in China. Several paper industries are surviving on bamboo forests. Ballarpur (Chandrapur, Maharstra) paper mills use bamboo for paper production.

Bamboo fiber has been used to make paper in China since early times. A high-quality, handmade paper is still produced in small quantities. Coarse bamboo paper is still used to make spirit money in many Chinese communities.

Bamboo pulps are mainly produced in China, Myanmar, Thailand, and India, and are used in printing and writing papers. The most common bamboo species used for paper are *Dendrocalamus asper* and *Bambusa blumeana*. It is also possible to make dissolving pulp from bamboo.

The average fiber length is similar to hardwoods, but the properties of bamboo pulp are closer to softwood pulps due to it having a very broad fiber length distribution. With the help of molecular tools, it is now possible to distinguish the superior fiber-yielding species/varieties even at juvenile stages of their growth, which can help in unadulterated merchandise production.

Weapons

Bamboo has often been used to construct weapons and is still incorporated in several Asian martial arts.

- A bamboo staff, sometimes with one end sharpened, is used in the Tamil martial art of *silambam*, a word derived from a term meaning "hill bamboo".
- Staves used in the Indian martial art of *gatka* are commonly made from bamboo, a material favoured for its light weight.
- A bamboo sword called a *shinai* is used in the Japanese martial art of *kendo*.
- Bamboo is used for crafting the bows, called *yumi*, and arrows used in the Japanese martial art *kyudo*.
- Bamboo is sometimes used to craft the limbs of the longbow and recurve bow used in traditional archery, and to make superior weapons for bowhunting and target archery.
- The first gunpowder-based weapons, such as the fire lance, were made of bamboo.
- Bamboo was apparently used in East and South Asia as a means of torture.

Musical Instruments

Other Uses

Tukir, cooking in bamboo in East Timor

Bamboo has traditionally been used to make a wide range of everyday utensils, particularly in Japan, where archaeological excavations have uncovered bamboo baskets dating to the Late Jomon period (2000–1000 BC). Bamboo has a long history of use in Asian furniture. Chinese bamboo furniture is a distinct style based on a millennia-long tradition. Several manufacturers offer bamboo bicycles, surfboards, snowboards, and skateboards. Due to its flexibility, bamboo is also used to make fishing rods. The split cane rod is especially prized for fly fishing. Bamboo has been traditionally used in Malaysia as a firecracker called a *meriam buluh* or bamboo cannon. Four-foot-long sections of bamboo are cut, and a mixture of water and calcium carbide are introduced. The resulting acetylene gas is ignited with a stick, producing a loud bang. Bamboo can be used in water desalination. A bamboo filter is used to remove the salt from seawater.

Food is cooked in East Timor in bamboo in fire. This is called *tukir*.

Bamboo is used for mussels breeding and propagation (Abucay, Bataan, Philippines).

Many minority groups in remote areas that have water access in Asia use bamboo that is 3–5 years old to make rafts. They use 8 to 12 poles, 6–7 m (20 to 24 ft) long, laid together side by side to a width of about 1 m (3.3 ft). Once the poles are lined up together, they cut a hole crosswise through the poles at each end and use a small bamboo pole pushed through that hole like a screw to hold all the long bamboo poles together. Floating houses use whole bamboo stalks tied together in a big bunch to support the house floating in the water. Bamboo is also used to make eating utensils such as chopsticks, trays, and tea scoops.

The Song Dynasty (960–1279 AD) Chinese scientist and polymath Shen Kuo (1031–1095) used the evidence of underground petrified bamboo found in the dry northern climate of Yan'an, Shanbei region, Shaanxi province to support his geological theory of gradual climate change.

Symbolism and Culture

Bamboo's long life makes it a Chinese symbol of uprightness, while in India it is a symbol of friendship. The rarity of its blossoming has led to the flowers' being regarded as a sign of impending famine. This may be due to rats feeding upon the profusion of flowers, then multiplying and destroying a large part of the local food supply. The most recent flowering began in May 2006. Bamboo is said to bloom in this manner only about every 50 years.

Bamboo, by Xu Wei, Ming Dynasty.

In Chinese culture, the bamboo, plum blossom, orchid, and chrysanthemum (often known as *méi lán zhú jú* 梅兰竹菊) are collectively referred to as the Four Gentlemen. These four plants also represent the four seasons and, in Confucian ideology, four aspects of the *junzi* ("prince" or "noble one"). The pine (*sōng* 松), the bamboo (*zhú* 竹), and the plum blossom (*méi* 梅) are also admired for their perseverance under harsh conditions, and are together known as the "Three Friends of Winter" (岁寒三友 *suìhán sāny* ǒ *u*) in Chinese culture. The "Three Friends of Winter" is traditionally used as a system of ranking in Japan, for example in sushi sets or accommodations at a traditional *ryokan*. Pine (*matsu* 松) is of the first rank, bamboo (*také* 竹) is of second rank, and plum (*ume* 梅) is of the third.

The Bozo ethnic group of West Africa take their name from the Bambara phrase *bo-so*, which means "bamboo house". Bamboo is also the national plant of St. Lucia.

Bamboo, Noble and Useful

Bamboo, one of the "Four Gentlemen" (bamboo, orchid, plum blossom and chrysanthemum), plays such an important role in traditional Chinese culture that it is even regarded as a behavior model of the gentleman. As bamboo has features such as uprightness, tenacity, and hollow heart, people endow bamboo with integrity, elegance, and plainness, though it is not physically strong. Countless poems praising bamboo written by ancient Chinese poets are actually metaphorically about people who exhibited these characteristics. According to laws, an ancient poet, Bai Juyi (772–846), thought that to be a gentleman, a man does not need to be physically strong, but he must be mentally strong, upright, and perseverant. Just as a bamboo is hollow-hearted, he should open his heart to accept anything of benefit and never have arrogance or prejudice. Bamboo is not only a symbol of a gentleman, but also plays an important role in Buddhism, which was introduced into China in the first century. As canons of Buddhism forbids cruelty to animals, flesh and egg were not allowed in the diet. The tender bamboo shoot (*s* ǔ *n*笋 in Chinese) thus became a nutritious alternative. Preparation methods developed over thousands of years have come to be incorporated into Asian cuisines, especially for monks. A Buddhist monk, Zan Ning, wrote a manual of the bamboo shoot called "*S* ŭ *n P* ŭ 筍譜" offering descriptions and recipes for many kinds of bamboo shoots. Bamboo shoot has always been a traditional dish on the Chinese dinner table, especially in southern China. In ancient times, those who could afford a big house with a yard would plant bamboo in their garden.

Photo of carved Chinese bamboo wall vase. 1918. Brooklyn Museum Archives, Goodyear Archival Collection.

In Japan, a bamboo forest sometimes surrounds a Shinto shrine as part of a sacred barrier against evil. Many Buddhist temples also have bamboo groves.

A cylindrical bamboo brush holder or holder of poems on scrolls, created by Zhang Xihuang in the 17th century, late Ming or early Qing Dynasty – in the calligraphy of Zhang's style, the poem *Returning to My Farm in the Field* by the fourth-century poet Tao Yuanming is incised on the holder.

Bamboo-style barred window in Lin An Tai Historical House, Taipei

Bamboo plays an important part of the culture of Vietnam. Bamboo symbolizes the spirit of Vovinam (a Vietnamese martial arts): *cương nhu phối triển* (coordination between hard and soft (martial arts)). Bamboo also symbolizes the Vietnamese hometown and Vietnamese soul: the gentlemanlike, straightforwardness, hard working, optimism, unity, and adaptability. A Vietnamese proverb says, "When the bamboo is old, the bamboo sprouts appear", the meaning being Vietnam will never be annihilated; if the previous generation dies, the children take their place. Therefore, the Vietnam nation and Vietnamese value will be maintained and developed eternally. Traditional Vietnamese villages are surrounded by thick bamboo hedges (*lũy tre*).

In Mythology

Several Asian cultures, including that of the Andaman Islands, believe humanity emerged from a bamboo stem.

In Philippine mythology, one of the more famous creation accounts tells of the first man, *Malakás* ("Strong"), and the first woman, *Maganda* ("Beautiful"), each emerged from one half of a split bamboo stem on an island formed after the battle between Sky and Ocean. In Malaysia, a similar story includes a man who dreams of a beautiful woman while sleeping under a bamboo plant; he wakes up and breaks the bamboo stem, discovering the woman inside. The Japanese folktale "Tale of the Bamboo Cutter" (*Taketori Monogatari*) tells of a princess from the Moon emerging from a shining bamboo section. Hawaiian bamboo *('ohe)* is a *kinolau* or body form of the Polynesian creator god Kāne.

A bamboo cane is also the weapon of Vietnamese legendary hero, Saint Giong, who had grown up immediately and magically since the age of three because of his wish to liberate his land from Ân invaders. An ancient Vietnamese legend (The Hundred-knot Bamboo Tree) tells of a poor, young farmer who fell in love with his landlord's beautiful daughter. The farmer asked the landlord for his daughter's hand in marriage, but the proud landlord would not allow her to be bound in marriage to a poor farmer. The landlord decided to foil the marriage with an impossible deal; the farmer must bring him a "bamboo tree of 100 nodes". But Gautama Buddha (*Bụt*) appeared to the farmer and told him that such a tree could be made from 100 nodes from several different trees.

Bụt gave to him four magic words to attach the many nodes of bamboo: *Khắc nhập, khắc xuất*, which means "joined together immediately, fell apart immediately". The triumphant farmer returned to the landlord and demanded his daughter. Curious to see such a long bamboo, the landlord was magically joined to the bamboo when he touched it, as the young farmer said the first two magic words. The story ends with the happy marriage of the farmer and the landlord's daughter after the landlord agreed to the marriage and asked to be separated from the bamboo.

In a Chinese legend, the Emperor Yao gave two of his daughters to the future Emperor Shun as a test for his potential to rule. Shun passed the test of being able to run his household with the two emperor's daughters as wives, and thus Yao made Shun his successor, bypassing his unworthy son. Later, Shun drowned in the Xiang River. The tears his two bereaved wives let fall upon the bamboos growing there explains the origin of spotted bamboo. The two women later became goddesses.

According to Maranaw

- Buntong or Bentong (e.i Palaw ah Bentong that means Mountain of Bamboos)
- Tringen *Tring*

References

- Joyce, Edward; revised and expanded by Alan Peters (1987). Encyclopedia of Furniture Making. New York: Sterling Publishing Co. ISBN 0-8069-6440-5.
- Soroka, W (2008). Illustrated Glossary of Packaging Terms. Institute of Packaging Professionals. p. 81. ISBN 1-930268-27-0.
- MacDonald, Nancy (2013). Woodworking (2nd ed.). Cengage Learning. pp. 782–784. ISBN 9781285700502. Retrieved 12 November 2015.
- "Wood waste: A short review of recent research" (PDF). Department for Environment, Food and Rural Affairs. July 2012. Retrieved 5 July 2016
- "Factors affecting distribution of borate to protect building envelope components from biodegradation" (PDF). Tspace.library.utoronto.ca. Retrieved 2015-08-27.
- "Forest Research - Social, cultural and economic values of contemporary non-timber forest products: Wild Harvests". Forestry.gov.uk. Retrieved 2013-11-21.
- Iggesund Paperboard (2008). "Product Catalogue: General Technical Information". Archived from the original on July 13, 2011. Retrieved 2009-07-18.
- Dr. Peter Ince (1999–2000). "Paper, Paperboard and Woodpulp – Production, Consumption and Trade ECE/FAO Forest Products Annual Market Review" (PDF). Archived from the original (PDF) on June 5, 2011. Retrieved 2009-07-18.
- Metso Corporation's press release (2011). "Metso to supply coated board line to International Paper & Sun Cartonboard in China". Retrieved 2011-11-04.
- Federal Institute for Risk Assessment (Germany). "Migration of mineral oil from packaging materials to foodstuffs (BfR Opinion No. 008/2010 of 2009-12-09)" (PDF). Retrieved 2010-04-21.
- Bundesinstitut für Risikobewertung. "Übergänge von Mineralöl aus Verpackungsmaterialien auf Lebensmittel" (PDF). Retrieved 2010-04-21.

6

Methods of Data Collection

Forestry requires systematic data collection on tree-felling, afforestation as well as the occurrence of natural hazards such as landslides and wildfires. Some methods of data collection include basal area, analog forestry and geographic information systems. The chapter serves as a source to understand the major categories related to forestry.

Climax Community

In ecology, climax community, or climatic climax community, is a historic term that expressed a biological community of plants, animals, and fungi which, through the process of ecological succession the development of vegetation in an area over time, had reached a steady state. This equilibrium was thought to occur because the climax community is composed of species best adapted to average conditions in that area. The term is sometimes also applied in soil development. Nevertheless, it has been found that a "steady state" is more apparent than real, particularly if long-enough periods of time are taken into consideration. Notwithstanding, it remains a useful concept.

The Daintree Rainforest in Queensland, Australia is an example of a climax forest ecosystem.

The idea of a single climatic climax, which is defined in relation to regional Climate, originated with Frederic Clements in the early 1900s. The first analysis of succession as leading to something like a climax was written by Henry Cowles in 1899, but it was Clements who used the term "climax" to describe the idealized endpoint of succession.

Frederic Clements's Use of "Climax"

Clements described the successional development of an ecological communities comparable to the ontogenetic development of individual organisms. Clements suggested only comparisons to very simple organisms. Later ecologists developed this idea that the ecological community is a "superorganism" and even sometimes claimed that communities could be homologous to complex organisms and sought to define a single climax-type for each area. The English botanist Arthur Tansley developed this idea with the "polyclimax" — multiple steady-state end-points, determined by edaphic factors, in a given climatic zone. Clements had called these end-points other terms, not climaxes, and had thought they were not stable, because by definition climax vegetation is best-adapted to the climate of a given area. Henry Gleason's early challenges to Clements's organism simile, and other of his strategies for describing vegetation, were largely disregarded for several decades until substantially vindicated by research in the 1950s and 1960s (below). Meanwhile, climax theory was deeply incorporated in both theoretical ecology and in vegetation management. Clements's terms such as pre-climax, post-climax, plagioclimax and disclimax continued to be used to describe the many communities which persist in states that diverge from the climax ideal for a particular area.

Though the views are sometimes attributed to him, Clements never argued that climax communities must always occur, or that the different species in an ecological community are tightly integrated physiologically, or that plant communities have sharp boundaries in time or space. Rather, he employed the idea of a climax community — of the form of vegetation best adapted to some idealized set of environmental conditions — as a conceptual starting point for describing the vegetation in a given area. There are good reasons to believe that the species best adapted to some conditions might appear there, when those conditions occur. But much of Clements's work was devoted to characterizing what happens when those ideal conditions do not occur. In those circumstances, vegetation other than the ideal climax will often occur instead. But those different kinds of vegetation can still be described as deviations from the climax ideal. Therefore, Clements developed a very large vocabulary of theoretical terms describing the various possible causes of vegetation, and various non-climax states vegetation adopts as a consequence. His method of dealing with ecological complexity was to define an ideal form of vegetation — the climax community — and describe other forms of vegetation as deviations from that ideal.

Discomfort with Climax Theory

Support among ecologists for the climax theory declined, because they found the theory with its many coined terms difficult to apply, because they were dissatisfied how it compared to observed individual organisms, and because better theories developed.

Although Clements recognized that vegetation follows gradients rather than being tightly bound, his rhetorical comparisons of ecological communities to organisms fostered the impression that communities, including the climax, have distinct edges in space and time. Yet Robert Whittaker's research demonstrated plant species distribute themselves along nutrient and other environmental gradients. Many ecologists saw this as a major reason to stop using the climax concept.

More recent palynological studies showed that modern species assemblages are ephemeral; vegetation in eastern North America since the last glacial maximum has consisted of several different species assemblages, many of which have no analogues in modern "climax" communities. That would mean, at least, that the climax types for those areas could not be stable to the degree Clements believed they were.

Ultimately, even if succession tends towards a steady state, the time required to achieve this state is unrealistically long; in most cases, external disturbances and environmental change occur so frequently that the realization of a climax community is unlikely, and therefore it has come to be regarded as a less useful concept. Long-term vegetation dynamics are now more often characterized as resulting from the action of stochastic factors.

Continuing Usage of "Climax"

Despite the overall abandonment of climax theory, during the 1990s use of climax concepts again became more popular among some theoretical ecologists. Many authors and nature-enthusiasts continue to use the term "climax" in a diluted form to refer to what might otherwise be called mature or old-growth communities. The term "climax" has also been adopted as description for a late successional stage for marine macroinvertebrate communities.

Additionally, some contemporary ecologists still use the term "disclimax" to describe an ecosystem been dominated by invasive species that competitively prevent the re-introduction of once native species. This concept borrows from Clement's earliest interpretation of climax as referring to an ecosystem that is resistant to colonization by outside species. The term disclimax was used in-context by Clements (1936), and despite being an anthropogenic phenomenon which prevents the facilitation and succession to a true climax community, it is one of the only examples of climax that can be observed in nature.

Basal Area

Basal area is the area of a given section of land that is occupied by the cross-section of tree trunks and stems at the base. The term is used in forest management and forest ecology.

In most countries, this is usually a measurement taken at the diameter at breast height (1.3m or 4.5 ft) of a tree above the ground and includes the complete diameter of every tree, including the bark. Measurements are usually made for a plot and this is then scaled up for 1 hectare of land for comparison purposes to examine a forest's productivity and growth rate.

To estimate a tree's basal area BA, use the tree's diameter at breast height DBH in inches with the following formula:

$$BA = \frac{\pi \times (DBH / 2)^2}{144}$$

Note: The factor of 144 is there to convert from Sq Inches to Sq Feet)

This formula simplifies to: $BA = 0.005454 \times DBH^2$

The result will be in ft^2.

For the DBH in cm use: $BA = 0.00007854 \times DBH^2$

The result will be in m^2.

The basal area of a forest stand can be found by adding the basal areas (as calculated above) of all of the trees in an area and dividing by the area of land in which the trees were measured. Basal area is generally expressed as ft^2/acre or m^2/ha.

A wedge prism can be used to quickly estimate the basal area per hectare. To find basal area using this method, simply multiply your BAF (Basal Area Factor) by the number of "in" trees in your variable radius plot. The BAF will vary based on the prism used, common BAFs include 5/8/10, and all "in" trees are those trees, when viewed through your prism from plot centre, that appear to be in-line with the standing tree on the outside of the prism.

Worked Example

If you carried out a survey using Angle Count Sampling (wedge prism) and you selected a Basal Area Factor (BAF) of 4. And you measured your first tree and its Diameter Breast Height (DBH) was 14 cm. Then the standard way of calculating how much of 1ha was covered by tree area (scaling up from that tree to the hectare) would be:

$4/((DBH+0.5)^2 \text{ X } \pi/4))$

- So this would be 4 - this is the BAF selected for that sampling technique.
- DBH would be 14 - this uses an assumed diameter, when actually used is the radius perpendicular to the tangent line.
- The + 0.5 allows under and over measurement to be accounted for.
- The $\pi/4$ converts the rest to the area.

Lastly to see the figure in m^2 it needs to be multiplied by 10 000. IN this case this means in every Ha there is 242m^2 of tree area according to this sampled tree being taken as representative of all the unmeasured trees.

Fixed Area Plot

It would also be possible to survey the trees in a Fixed Area Plot (FAP). Also called a Fixed Radius Plot. In the case that this plot was 100m^2. Then the formula would be

$(DBH+0.5)^2 \text{X } \pi/4$

Analog Forestry

Analogue forestry is a system of planned, managed forests, primarily employed in tropical or subtropical areas. The forests are designed to mimic the function and ecology of the preexisting climax vegetation for the area, and are also designed to provide economic benefits.

Analogue forestry draws design input not only from traditional models but also from the natural forest successional dynamics. When an ecosystem is designed to be analogous to the indigenous climax state, the efficiency and dynamics of the natural processes can be replicated. These quasi-natural forests are designed to mimic the structural and functional aspects of indigenous forests and are referred to as analog forests. In addition to their ecological characteristics, analog forests are also designed to provide economic benefits. However, it is not until all the ecological requirements of the location are satisfied that economic values of species are considered. Therefore, an analog forest may comprise natural and exotic species in any proportion, the contribution to structure and function being the overriding factor that determines its use.

It arose in Sri Lanka around 1981 as an alternative to monocultures of Pinus and Eucalyptus and has spread to India, Vietnam, Philippines, Australia, Peru, Ecuador, Colombia, Brazil, Costa Rica, Dominican Republic, Honduras, Mexico, Canada, Kenya and Zimbabwe at present.

The International Analog Forestry Network (IAFN) is currently hosted in Costa Rica.

Analogue forestry is a system of silviculture that seeks to design, plan and manage ecosystems dominated by woody perennials. It has been primarily employed in tropical or subtropical areas, but can be used in temperate areas too. The design seeks to mimic the architectural structure and ecological function of the preexisting climax vegetation of the area, and can be designed to provide economic, social and environmental benefits.

Analogue forestry always considers the inclusion of non-crop native species to be considered in the design and thus addresses the inclusion of native biodiversity in the design. As analog forestry also requires the inclusion of long-lived species of trees in the design it has the capacity to sequester carbon for a longer time than plantation forestry. Analogue forestry has the potential to produce very high values of photosynthetic biomass as its design calls for the inclusion of all the growth forms that occupy the three-dimensional space of the mature indigenous forest. By including many species of crops in one area analog forestry helps spread the risk of market failure on a single crop.

Forest Informatics

Forest informatics is the combined science of Forestry and informatics, with a special emphasis on collection, management, and processing of data, information and knowledge, and the incorporation of informatic concepts and theories specific to enrich forest management and forest science; it has a similar relationship to library science and information science.

It is an interdisciplinary science primarily concerned with the collection, classification, manipulation, storage, retrieval and dissemination of information. Information, in this context, includes both human and machine readable documents. Examples of human readable documents include maps, field data sheets, operational schedules, and long term asset management plans with narrative text. Machine readable documents include files for geographic information systems (GIS), Global Positioning Systems (GPS), and other applications like spreadsheets, and relational database management systems.

As in management science, Forest Informatics uses decision support systems, mathematical modeling, statistics, and algorithms from engineering, operations research, computer science, and artificial intelligence to support decision-making activities. Common forestry problems include harvest scheduling, model fitting, optimal sampling, remote sensing, crew assignment, image classification, treatment timing, and log bucking problems, many of which can be formulated as optimization problems (e.g. generalized assignment problem, traveling salesman problem, knapsack problem, job shop scheduling, and vehicle routing problems). The practice includes information processing and the engineering of information systems, decision support systems, geographic information systems, and global positioning systems. The research field includes studies the structure, algorithms, behavior, and interactions of natural and artificial systems that store, process, access and communicate information about forested ecosystems.

History

In 1970, J. G. Grevatt wrote an article titled, "Management Information and Computers in Forestry." In the article, the author describes and discusses different dimensions of management information (i.e. operation, expenditure, location, and time) including the nature of management information and decisions, management information in forestry, the management information system itself, the application of computers, the structure of a computer based system, comparisons between clerical and computer systems, and the impact on the field manager. The author concludes that the use of computers to process management data may be justified on grounds of cost and improved information in organizations of a critical size.

At the time of that article, computers, databases, and geographic information systems were still in their infancy and tools like the Global Positioning Systems of today were yet invented. Management database systems for business were more prevalent. Over the next 30 years, computers became more powerful, smaller, and less expensive. Relational database management systems had become commonplace in business, interrogating the computer system had become standardized with languages like SQL, and faster networks for data and information integration have become highly integrated. In that time, geographic information systems that could run on desktop computers and could be customized for various tasks were also developed, but as separate systems.

Within the last 10 years, specialized fields of study at the university level are offered at the several forestry schools where students learn the principles of quantification, modeling, descriptive and predictive analyses of natural resources attributes needed for sound management of forested ecosystems.

At the small forester practitioner level, more software "back-end" programs have become available to model likely forest growth outcomes based on treatment prescriptions. These are provided by private businesses, such as Assisisoft, as well as government agencies, such as the U.S. Forestry Service's NED system. The basic functions of the NED system is commonly used among American consulting foresters, as it is free, although a small proportion of that group uses the full modeling capabilities of the software. With the prevalence of smartphone and tablet access, the computing power available in the woods is now much higher than it was just a few years ago when PocketPC-based systems were prevalent. Microsoft announced it will deprecate the PocketPC platform in March 2013. New "apps," such as Forest Metrix are now becoming available for foresters and timber cruisers to employ their devices for data collection for later export into more sophisticated software.

Software specifically devoted to analyzing management decisions for forested ecosystems have been developed, and used in several large scale planning projects. For example, the Ecosystem Management Decision Support (EMDS) system is an application framework for knowledge-based decision support of ecological analysis and planning. Open source software solutions have also become more widely accepted as well, as is seen in the expansion of ecological extensions for statistical tools like R. A recent example would be the book written by Andrew Robinson and Jeff D. Hamann about using R for forest analytics.

In 2006, the United Nations declared 2011 to be International Year of Forests.

Forest Informatics, Inc. has developed a postgresql template, a set of software agents, and a collection of reports, maps, and data feeds. The application uses an intelligent agent architecture to preemptively generate possible strategic, tactical, and operational solutions for forest managers.

Forest Inventory

Forest inventory is the systematic collection of data and forest information for assessment or analysis. An estimate of the value and possible uses of timber is an important part of the broader information required to sustain ecosystems. When taking forest inventory the following are important things to measure and note: species, diameter at breast height (DBH), height, site quality, age, and defects. From the data collected one can calculate the number of trees per acre, the basal area, the volume of trees in an area, and the value of the timber. Inventories can be done for other reasons than just calculating the value. A forest can be cruised to visually assess timber and determine potential fire hazards and the risk of fire. The results of this type of inventory can be used in preventative actions and also awareness. Wildlife surveys can be undertaken in conjunction with timber inventory to determine the number and type of wildlife within a forest. The aim of the statistical forest inventory is to provide comprehensive information about the state and dynamics of forests for strategic and management planning. Merely looking at the forest for assessment is called taxation.

History

Surveying and taking inventory of trees originated in Europe in the late 18th century out of a fear that wood (the main source of fuel) would run out. The first information was organized into maps used to plan out usage. In the early 19th century forest harvesters estimated the volume and dispersal of trees within smaller forests with their eyes. More diverse and larger forests were divided into smaller sections of similar type trees that were individually estimated by visual inspection. These estimates were related together to figure out the entire forest's available resources. As the 19th century progressed so did the measurement techniques. New relationships between diameter, height, and volume were discovered and exploited. These newfound relationships allowed for a more accurate assessment of wood types and yields of much larger forests. By 1891, these surveys were conducted through sample-based methods involving statistical averages and more sophisticated measuring devices were implemented. In the 20th century, the statistical method of sampling had become well established and commonly used. Further developments, such as unequal probability sampling, arose. As the 20th century progressed, an understanding of co-efficients of error became clearer and the new technology of computers combined with the availability of aerial as well as satellite photography, further refined the process. Laser Scanning both terrestrially and aerially are now used alongside more manual methods. As a result, sampling accuracy and assessment values became more accurate and allowed for modern practices to arise.

A forest inventory does not only record the trees height, DBH and number for tree yield calculations. But it also records the conditions of the forest. So this would include, geology, site conditions, tree health and other forest factors.

Timber Cruise

A timber cruise is a sample measurement of a stand used to estimate the amount of standing timber that the forest contains. These measurements are collected at sample locations called plots, quadrants, or strips. Each of these individual sample areas is one observation in a series of observations called a sample. These sample areas are generally laid out in some random fashion usually in the form of a line plot survey. Depending on the size of the plot and the number of plots measured, the data gathered from these plots can then be manipulated to achieve varying levels of certainty for an estimate that can be applied to the entire timber stand. This estimate of stand conditions, species composition, volume and other measured attributes of a forest system can then be used for various purposes. For example, in British Columbia the sale of Crown timber is a business proposition and both the buyer and the Ministry of Forests and Range (seller) must know the quantity and the quality of timber being sold. Our satellite provides the essential data and information for determining stumpage rates, for establishing conditions of sale and for planning of the logging operations by the licensee. Generally a timber cruise includes measurements or estimates of timber volume by forest product sort (and sometimes grade), log defect, and log lengths, whether the estimates are made in the field or using computer software.

Stranded Examinations

Example of hardware equipment for forest inventories: GPS and laser rangefinder for mapping connected to a field rugged computer.

Before, the most common type of inventory is one that uses a random sampling technique which groups similar forests into one category based on age, stand structure, species, and location. The next step would be to begin from a random place and measure circular plots that are equidistant from one another. Inventory undertaken in conjunction with wildlife surveys has a regularized distance between plots. When taking inventory, a decision has to be made about which types of plots to measure. There are several different types of plots. The two most common types are *fixed radius* and *variable radius*, also known as *prism radius*. In a fixed radius plot, the forester finds the center of a plot and every tree within a certain fixed distance away from that point is measured. Variable radius plots are used more for inventory of volume. During this method, an angle is created and projected from the plot center and all trees that are larger than the projected right angle are measured with a ruler called a Tunagmetor .

Types of Sample Plots

Fixed Area Plot

Fixed area plot sample measurements are taken so that they are a fraction of the entire timber stand. This means that the numbers are all proportional to the actual stand values and that by multiplying by the correct corresponding value you can obtain the actual tract values. These plots are taken randomly so that each sample point has an equal probability of being included in the random sample. Commonly a 100m² plot is taken.

Variable Size Plots

A variable size plot is more dependent of the size of the trees. The tract is measured on a series of points and the trees are tallied for being in or out depending on their size and location relevant to the plot center. Usually an angle gauge, wedge prism or Relascope are used to gather data for this type of plot. This allows for a very quick estimate of the volume and species of a given tract.

Transects

Transects are arbitrarily determined lines (to prevent sampling bias) through a stand employed as a linear form of sample plot. They are sometimes referred to as "strip lines."

Plot Selection

Plots are samples of the forest being inventoried and so are selected according to what is looked for.

Simple Random Sampling

A computer or calculator random number generator is used to assign plots to be sampled. Here random means an equal chance of any plot being selected out of all of the plots available. It does not mean haphazard. Often it is modified to avoid sampling roads, ensure coverage of unsampled areas and for logistics of actually getting to the plots.

Systematic Sampling

Commonly this is done by a random point and then laying a grid over a map of the area to be sampled. This grid will have preassigned plot areas to be sampled. It means more efficient logistics and removes some of the human bias that may be there with simple random sampling.

Systematic Stratified Sampling

There is some HAPPY grouping, for example by age classes or soil characteristics or slope elevation. And then plots are chosen from each grouping by another sampling technique. It requires some knowledge of the land first and also trust that the groupings have been done properly. In forestry it may be done to separate plantation areas from mixed forest for example and reduce the amount of sampling time needed.

Systematic Clustered Sampling

When it is not possible to make strata for stratified sampling, there may be some knowledge about the forest where it can be said that small groupings are possible. These small groupings of plots if they are near to each other form a cluster. These clusters are then randomly sampled with the belief that they are representing the actual mix of the forest. As they are close to each other there is less waking needed and so it is more efficient.

Timber Metrics

The amount of standing timber that a forest contains is determined from:

- Age class. This is a misnomer as in German it is Wuchsklass (Growth class) and properly should be the size of the tree (Size class), which may be limited by shade over it and not necessarily the biological and thus physiological age of the tree. This distinction is important if tree growth over time is expected by an owner or forester to produce timber, as a small old tree will grow differently from a small young tree. Commonly these age classes are split into: Seedling, Sapling, Pole, Mature Tree (subdivided into Weak wood, Middle wood and Strong wood stages), Old / Scenescant Tree. Sometimes it is called size class or a cohort. There are differences between countries and forests.
- Basal area – defines the area of a given section of land that is occupied by the cross-section of tree trunks and stems at their base
- Diameter at breast height (DBH) – measurement of a tree's girth standardized with different countries having different standards they are often at 1.3 meters (about 4.5 feet) above the ground
- Form factor – the shape of the tree, based on recorded trees and commonly then given for calculating tree volumes for a given species. It is usually related to DBH or age class. It is distinct from taper. So it can be conical or paraboloid for example.
- Girard form class – an expression of tree taper calculated as the ratio of diameter inside the bark at 16 feet above ground to that outside bark at DBH, primary expression of tree form used in the United States
- Quadratic mean diameter – diameter of the tree that coordinates to the stand's basal area
- Site index – a species specific measure of site productivity and management options, reported as the height of dominant and co-dominant trees (site trees)in a stand at a base age such as 25, 50 and 100 years
- Tree taper – the degree to which a tree's stem or bole decreases in diameter as a function of height above ground. So it can be sharp or gradual.

Volume Estimation

- Stocking – a quantitative measure of the area occupied by trees relative to an optimum or desired level of density, which will vary according to management purpose
- Stand Density Index – a measure of the stocking of a stand of trees based on the number of trees per unit area and DBH of the tree of average basal area, based on historical records and is particular to site type and species
- Volume table – a chart based on volume equations that uses correlations between certain aspects of a tree to estimate the standing volume
- Stand density management diagram – model that uses current stand density to project future stand composition

Volume can be calculated from the metrics recorded in a plot sample. For example, if a tree was measured to be 20m tall and with a DBH of 19 cm using previous measured tree data a volume could be approximated according to species. Such a table has been constructed by Josef Pollanschütz in Austria.

Volume of tree = BA X h x f pollanschutz

So f pollanschütz would be derived from the table and is properly called the Form Factor.

To scale this up to a hectare level the result would have to be multiplied by the number of trees of that size. This is called the blow up factor.

Tools Used in Inventory

- Biltmore stick – utilizes ocular trigonometry to quickly measure diameter and height

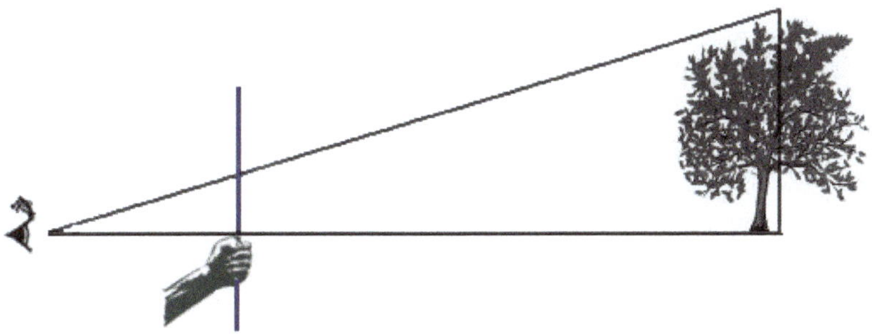

Figure demonstrating the ocular trigonometric principles behind the Biltmore stick.

- Diameter tape – cloth or metal tape that is wrapped around the bole, scaled to diameter
- Caliper – two prongs connected to a measuring tape are placed around the most average part of the bole to determine diameter
- Relascope – multiple-use tool that is able to find tree height, basal area, and tree diameter anywhere along the bole
- Clinometer – common tool used to measure changes in elevation and tree height

- Cruising rod – similar to a caliper, calculates the number of pieces of lumber yielded by a given piece of timber by measuring its diameter
- LASER Scanner used with computer software to calculate the metrics from the collected data by use of Lidar.
- Wedge prism – a small glass wedge that refracts light to create visible offsets in order to be able to choose which trees at a sampling point should be included in the sample.
- Data collector – an electronic device used to quickly enter sample data, geo-locate the data, and, in more modern times, to also access reference, web and historic materials while timber cruising.
- Increment borer – a device used to retrieve a cylindrical sample of wood material orthogonally from the stem while doing as little damage as possible to the remaining tissues.

In 2014, the Food and Agriculture Organization of the United Nations and partners, with the support of the Government of Finland, launched Open Foris – a set of open-source software tools that assist countries in gathering, producing and disseminating reliable information on the state of forest resources. The tools support the entire inventory lifecycle, from needs assessment, design, planning, field data collection and management, estimation analysis, and dissemination. Remote sensing image processing tools are included, as well as tools for international reporting for REDD+ MRV and FAO's Global Forest Resource Assessments.

Tree Allometry

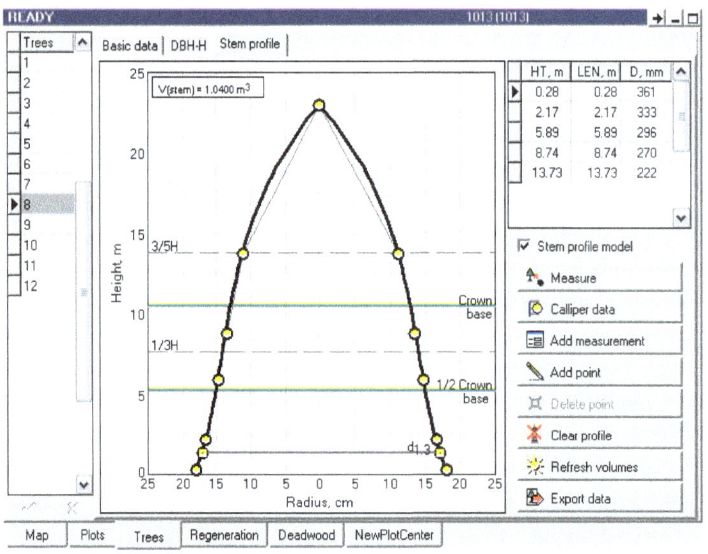

Stem profile measurement Electronic equipment (as Field-Map for example) is used for stem profile measurements and for measurements of profiles/projections of the crown. These key measurements are used for estimation of carbon sequestered in the plants.

Tree allometry establishes quantitative relations between some key characteristic dimensions of trees (usually fairly easy to measure) and other properties (often more difficult to assess). To the extent these statistical relations, established on the basis of detailed measurements on a small sample of typical trees, hold for other individuals, they permit extrapolations and estimations of a host of dendrometric quantities on the basis of a single (or at most a few) measurements.

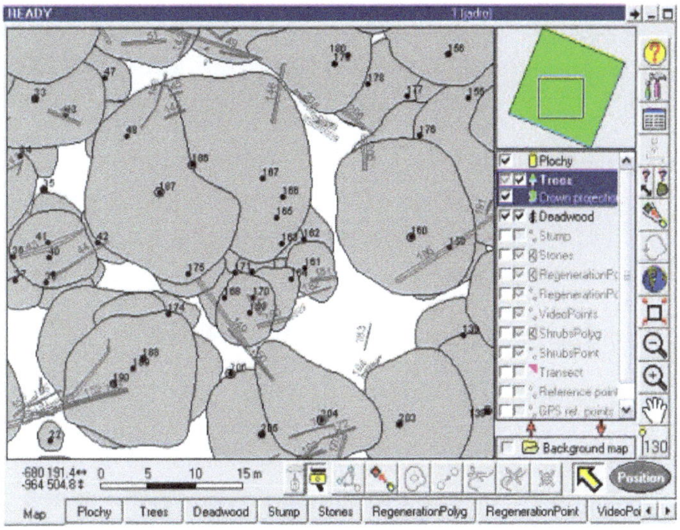

Forest structure measurement is needed for establishment of allometric equations.

The study of allometry is extremely important in dealing with measurements and data analysis in the practice of forestry. Allometry studies the relative size of organs or parts of organisms. Tree allometry narrows the definition to applications involving measurements of the growth or size of trees. Allometric relationships are often estimating difficult tree measurement, such as volume, from an easily measured attribute such as diameter at breast height (DBH).

The use of allometry is widespread in forestry and forest ecology. In order to develop an allometric relationship there must be a strong relationship and an ability to quantify this relationship between the parts of the subject measured and the other quantities of interest. Also when developing this equation one must play in factors which affect tree growth such as age, species, site location, etc. Once all these guidelines are met, one may attempt to develop an allometric equation.

Assessment of Forest Biomass and Carbon Stocks

In 2013, the Food and Agriculture Organization of the United Nations launched GlobAllomeTree, a web-based platform designed to improve global access to tree allometric equations and support forest and climate-change project developers, researchers, scientists and foresters to assess forest volume and biomass, and carbon stocks. Jointly developed by FAO, the French Research Centre CIRAD and Tuscia University of Italy, the GlobAllomeTree platform provides a consistent and harmonized database of tree and stand volume and biomass allometric equations; software to compare equations and assess variables of interests, such as volume, biomass and carbon stocks; access to scientific research information on allometric equations; and access to tutorials, manuals and documentation supporting the development and use of tree allometric equations.

In 2012, FAO and CIRAD published a manual for building tree volume and biomass allometric equations for students, technicians or researchers working to assess forest resources such as volume, biomass and carbon stocks for commercial, bioenergy or climate change mitigation purposes.

Methodology

First thing to do is select a group of some subject (for forestry: trees). Then measure several easily measured attributes such as DBH, height, species, etc. Graph the results and perform a regression analysis and transform some of the variables until a correct regression is found.

There are different tree species compositions in each region in the world and most of those regions have at least one equation that estimates tree volume from DBH. Research and the application of forest allometry have meshed over time to develop these quick equations to accurately estimate how much volume a particular forest stand holds.

The general allometric equation for mathematics and science is

$$Y = \beta X^{\alpha}$$

where "Y" is a biological variable (such as tree height or DBH), "β" is a proportionality coefficient, "α" is the scaling exponent (which is equal to the slope of the line when plotted on logarithmic coordinates), and "X" is some physical measure such as body volume or body mass(M). While α is often quite similar between very diverse organisms, β differs from species to species. Because the proportionality constant(β) and the scaling exponents(α) are often denoted using Greek letters, it is desirable to use β as the proportionality coefficient versus α, since α could be misread as the symbol for «proportional».

A well-known allometric equation relates metabolic rate to body mass: Y = βM 3/4.

In forestry the equation takes on many forms in order to represent relationships between the many various attributes of tree size and growth. Below is an example:

$$Y = b_o + b_1 X$$

Geographic Information System

A geographic information system or geographical information system (GIS) is a system designed to capture, store, manipulate, analyze, manage, and present all types of spatial or geographical data. The acronym GIS is sometimes used for geographic information science (GIScience) to refer to the academic discipline that studies geographic information systems and is a large domain within the broader academic discipline of geoinformatics. What goes beyond a GIS is a spatial data infrastructure, a concept that has no such restrictive boundaries.

In a general sense, the term describes any information system that integrates, stores, edits, analyzes, shares, and displays geographic information. GIS applications are tools that allow users to create interactive queries (user-created searches), analyze spatial information, edit data in maps, and present the results of all these operations. Geographic information science is the science underlying geographic concepts, applications, and systems.

GIS is a broad term that can refer to a number of different technologies, processes, and methods. It is attached to many operations and has many applications related to engineering, planning, management, transport/logistics, insurance, telecommunications, and business. For that reason, GIS and location intelligence applications can be the foundation for many location-enabled services that rely on analysis and visualization.

GIS can relate unrelated information by using location as the key index variable. Locations or extents in the Earth space–time may be recorded as dates/times of occurrence, and x, y, and z coordinates representing, longitude, latitude, and elevation, respectively. All Earth-based spatial–temporal location and extent references should, ideally, be relatable to one another and ultimately to a "real" physical location or extent. This key characteristic of GIS has begun to open new avenues of scientific inquiry.

History of Development

The first known use of the term "geographic information system" was by Roger Tomlinson in the year 1968 in his paper "A Geographic Information System for Regional Planning". Tomlinson is also acknowledged as the "father of GIS".

E. W. Gilbert's version (1958) of John Snow's 1855 map of the Soho cholera outbreak showing the clusters of cholera cases in the London epidemic of 1854

Previously, one of the first applications of spatial analysis in epidemiology is the 1832 "*Rapport sur la marche et les effets du choléra dans Paris et le département de la Seine*". The French geographer Charles Picquet represented the 48 districts of the city of Paris by halftone color gradient according to the number of deaths by cholera per 1,000 inhabitants. In 1854 John Snow determined the source of a cholera outbreak in London by marking points on a map depicting where the cholera victims lived, and connecting the cluster that he found with a nearby water source. This was one of the earliest successful uses of a geographic methodology in epidemiology. While the basic elements of topography and theme existed previously in cartography, the John Snow map was unique, using cartographic methods not only to depict but also to analyze clusters of geographically dependent phenomena.

The early 20th century saw the development of photozincography, which allowed maps to be split into layers, for example one layer for vegetation and another for water. This was particularly used for printing contours – drawing these was a labour-intensive task but having them on a separate layer meant they could be worked on without the other layers to confuse the draughtsman. This work was originally drawn on glass plates but later plastic film was introduced, with the advantages of being lighter, using less storage space and being less brittle, among others. When all the layers were finished, they were combined into one image using a large process camera. Once color printing came in, the layers idea was also used for creating separate printing plates for each color. While the use of layers much later became one of the main typical features of a contemporary GIS, the photographic process just described is not considered to be a GIS in itself – as the maps were just images with no database to link them to.

Computer hardware development spurred by nuclear weapon research led to general-purpose computer "mapping" applications by the early 1960s.

The year 1960 saw the development of the world's first true operational GIS in Ottawa, Ontario, Canada by the federal Department of Forestry and Rural Development. Developed by Dr. Roger Tomlinson, it was called the Canada Geographic Information System (CGIS) and was used to store, analyze, and manipulate data collected for the Canada Land Inventory – an effort to determine the land capability for rural Canada by mapping information about soils, agriculture, recreation, wildlife, waterfowl, forestry and land use at a scale of 1:50,000. A rating classification factor was also added to permit analysis.

CGIS was an improvement over "computer mapping" applications as it provided capabilities for overlay, measurement, and digitizing/scanning. It supported a national coordinate system that spanned the continent, coded lines as arcs having a true embedded topology and it stored the attribute and locational information in separate files. As a result of this, Tomlinson has become known as the "father of GIS", particularly for his use of overlays in promoting the spatial analysis of convergent geographic data.

CGIS lasted into the 1990s and built a large digital land resource database in Canada. It was developed as a mainframe-based system in support of federal and provincial resource planning and management. Its strength was continent-wide analysis of complex datasets. The CGIS was never available commercially.

In 1964 Howard T. Fisher formed the Laboratory for Computer Graphics and Spatial Analysis at the Harvard Graduate School of Design (LCGSA 1965–1991), where a number of important theoretical concepts in spatial data handling were developed, and which by the 1970s had distributed seminal software code and systems, such as SYMAP, GRID, and ODYSSEY – that served as sources for subsequent commercial development—to universities, research centers and corporations worldwide.

By the late 1970s two public domain GIS systems (MOSS and GRASS GIS) were in development, and by the early 1980s, M&S Computing (later Intergraph) along with Bentley Systems Incorporated for the CAD platform, Environmental Systems Research Institute (ESRI), CARIS (Computer Aided Resource Information System), MapInfo Corporation and ERDAS (Earth Resource Data Analysis System) emerged as commercial vendors of GIS software, successfully incorporating many of the CGIS features, combining the first generation approach to separation of spatial and attribute information with a second generation approach to organizing attribute data into database structures.

In 1986, Mapping Display and Analysis System (MIDAS), the first desktop GIS product emerged for the DOS operating system. This was renamed in 1990 to MapInfo for Windows when it was ported to the Microsoft Windows platform. This began the process of moving GIS from the research department into the business environment.

By the end of the 20th century, the rapid growth in various systems had been consolidated and standardized on relatively few platforms and users were beginning to explore viewing GIS data over the Internet, requiring data format and transfer standards. More recently, a growing number of free, open-source GIS packages run on a range of operating systems and can be customized to perform specific tasks. Increasingly geospatial data and mapping applications are being made available via the world wide web.

Several articles on the history of GIS have been published.

GIS Techniques and Technology

Modern GIS technologies use digital information, for which various digitized data creation methods are used. The most common method of data creation is digitization, where a hard copy map or survey plan is transferred into a digital medium through the use of a CAD program, and geo-referencing capabilities. With the wide availability of ortho-rectified imagery (from satellites, aircraft, Helikites and UAVs), heads-up digitizing is becoming the main avenue through which geographic data is extracted. Heads-up digitizing involves the tracing of geographic data directly on top of the aerial imagery instead of by the traditional method of tracing the geographic form on a separate digitizing tablet (heads-down digitizing).

Relating Information From Different Sources

GIS uses spatio-temporal (space-time) location as the key index variable for all other information. Just as a relational database containing text or numbers can relate many different tables using common key index variables, GIS can relate otherwise unrelated information by using location as the key index variable. The key is the location and/or extent in space-time.

Any variable that can be located spatially, and increasingly also temporally, can be referenced using a GIS. Locations or extents in Earth space–time may be recorded as dates/times of occurrence, and x, y, and z coordinates representing, longitude, latitude, and elevation, respectively. These GIS coordinates may represent other quantified systems of temporo-spatial reference (for example, film frame number, stream gage station, highway mile-marker, surveyor benchmark, building address, street intersection, entrance gate, water depth sounding, POS or CAD drawing origin/units). Units applied to recorded temporal-spatial data can vary widely (even when using exactly the same data, see map projections), but all Earth-based spatial–temporal location and extent references should, ideally, be relatable to one another and ultimately to a "real" physical location or extent in space–time.

Related by accurate spatial information, an incredible variety of real-world and projected past or future data can be analyzed, interpreted and represented. This key characteristic of GIS has begun to open new avenues of scientific inquiry into behaviors and patterns of real-world information that previously had not been systematically correlated.

GIS Uncertainties

GIS accuracy depends upon source data, and how it is encoded to be data referenced. Land surveyors have been able to provide a high level of positional accuracy utilizing the GPS-derived positions. High-resolution digital terrain and aerial imagery, powerful computers and Web technology are changing the quality, utility, and expectations of GIS to serve society on a grand scale, but nevertheless there are other source data that affect overall GIS accuracy like paper maps, though these may be of limited use in achieving the desired accuracy.

In developing a digital topographic database for a GIS, topographical maps are the main source, and aerial photography and satellite imagery are extra sources for collecting data and identifying attributes which can be mapped in layers over a location facsimile of scale. The scale of a map and geographical rendering area representation type are very important aspects since the information content depends mainly on the scale set and resulting locatability of the map's representations. In order to digitize a map, the map has to be checked within theoretical dimensions, then scanned into a raster format, and resulting raster data has to be given a theoretical dimension by a rubber sheeting/warping technology process.

A quantitative analysis of maps brings accuracy issues into focus. The electronic and other equipment used to make measurements for GIS is far more precise than the machines of conventional map analysis. All geographical data are inherently inaccurate, and these inaccuracies will propagate through GIS operations in ways that are difficult to predict.

Data Representation

GIS data represents real objects (such as roads, land use, elevation, trees, waterways, etc.) with digital data determining the mix. Real objects can be divided into two abstractions: discrete objects (e.g., a house) and continuous fields (such as rainfall amount, or elevations). Traditionally, there are two broad methods used to store data in a GIS for both kinds of abstractions mapping references: raster images and vector. Points, lines, and polygons are the stuff of mapped location attribute references. A new hybrid method of storing data is that of identifying point clouds, which combine three-dimensional points with RGB information at each point, returning a "3D color image". GIS thematic maps then are becoming more and more realistically visually descriptive of what they set out to show or determine.

Data Capture

Example of hardware for mapping (GPS and laser rangefinder) and data collection (rugged computer). The current trend for geographical information system (GIS) is that accurate mapping and data analysis are completed while in the field. Depicted hardware (field-map technology) is used mainly for forest inventories, monitoring and mapping.

Data capture—entering information into the system—consumes much of the time of GIS practitioners. There are a variety of methods used to enter data into a GIS where it is stored in a digital format.

Existing data printed on paper or PET film maps can be digitized or scanned to produce digital data. A digitizer produces vector data as an operator traces points, lines, and polygon boundaries from a map. Scanning a map results in raster data that could be further processed to produce vector data.

Survey data can be directly entered into a GIS from digital data collection systems on survey instruments using a technique called coordinate geometry (COGO). Positions from a global navigation satellite system (GNSS) like Global Positioning System can also be collected and then imported into a GIS. A current trend in data collection gives users the ability to utilize field computers with the ability to edit live data using wireless connections or disconnected editing sessions. This has been enhanced by the availability of low-cost mapping-grade GPS units with decimeter accuracy in real time. This eliminates the need to post process, import, and update the data in the office after fieldwork has been collected. This includes the ability to incorporate positions collected using a laser rangefinder. New technologies also allow users to create maps as well as analysis directly in the field, making projects more efficient and mapping more accurate.

Remotely sensed data also plays an important role in data collection and consist of sensors attached to a platform. Sensors include cameras, digital scanners and lidar, while platforms usually consist of aircraft and satellites. In England in the mid 1990s, hybrid kite/balloons called Helikites first pioneered the use of compact airborne digital cameras as airborne Geo-Information Systems. Aircraft measurement software, accurate to 0.4 mm was used to link the photographs and measure the ground. Helikites are inexpensive and gather more accurate data than aircraft. Helikites can be used over roads, railways and towns where UAVs are banned.

Recently with the development of miniature UAVs, aerial data collection is becoming possible with them. For example, the Aeryon Scout was used to map a 50-acre area with a Ground sample distance of 1 inch (2.54 cm) in only 12 minutes.

The majority of digital data currently comes from photo interpretation of aerial photographs. Soft-copy workstations are used to digitize features directly from stereo pairs of digital photographs. These systems allow data to be captured in two and three dimensions, with elevations measured directly from a stereo pair using principles of photogrammetry. Analog aerial photos must be scanned before being entered into a soft-copy system, for high-quality digital cameras this step is skipped.

Satellite remote sensing provides another important source of spatial data. Here satellites use different sensor packages to passively measure the reflectance from parts of the electromagnetic spectrum or radio waves that were sent out from an active sensor such as radar. Remote sensing collects raster data that can be further processed using different bands to identify objects and classes of interest, such as land cover.

When data is captured, the user should consider if the data should be captured with either a relative accuracy or absolute accuracy, since this could not only influence how information will be interpreted but also the cost of data capture.

After entering data into a GIS, the data usually requires editing, to remove errors, or further processing. For vector data it must be made "topologically correct" before it can be used for some advanced analysis. For example, in a road network, lines must connect with nodes at an intersection. Errors such as undershoots and overshoots must also be removed. For scanned maps, blemishes on the source map may need to be removed from the resulting raster. For example, a fleck of dirt might connect two lines that should not be connected.

Raster-to-vector Translation

Data restructuring can be performed by a GIS to convert data into different formats. For example, a GIS may be used to convert a satellite image map to a vector structure by generating lines around all cells with the same classification, while determining the cell spatial relationships, such as adjacency or inclusion.

More advanced data processing can occur with image processing, a technique developed in the late 1960s by NASA and the private sector to provide contrast enhancement, false color rendering and a variety of other techniques including use of two dimensional Fourier transforms. Since digital data is collected and stored in various ways, the two data sources may not be entirely compatible. So a GIS must be able to convert geographic data from one structure to another. In so doing, the implicit assumptions behind different ontologies and classifications require analysis. Object ontologies have gained increasing prominence as a consequence of object-oriented programming and sustained work by Barry Smith and co-workers.

Projections, Coordinate Systems, and Registration

The earth can be represented by various models, each of which may provide a different set of coordinates (e.g., latitude, longitude, elevation) for any given point on the Earth's surface. The simplest model is to assume the earth is a perfect sphere. As more measurements of the earth have accumulated, the models of the earth have become more sophisticated and more accurate. In fact, there are models called datums that apply to different areas of the earth to provide increased accuracy, like NAD83 for U.S. measurements, and the World Geodetic System for worldwide measurements.

Spatial Analysis with Geographical Information System (GIS)

GIS spatial analysis is a rapidly changing field, and GIS packages are increasingly including analytical tools as standard built-in facilities, as optional toolsets, as add-ins or 'analysts'. In many instances these are provided by the original software suppliers (commercial vendors or collaborative non commercial development teams), while in other cases facilities have been developed and are provided by third parties. Furthermore, many products offer software development kits (SDKs), programming languages and language support, scripting facilities and/or special interfaces for developing one's own analytical tools or variants. The website "Geospatial Analysis" and associated book/ebook attempt to provide a reasonably comprehensive guide to the subject. The increased availability has created a new dimension to business intelligence termed "spatial intelligence" which, when openly delivered via intranet, democratizes access to geographic and social network data. Geospatial intelligence, based on GIS spatial analysis, has also become a key element for security. GIS as a whole can be described as conversion to a vectorial representation or to any other digitisation process.

Slope and Aspect

Slope can be defined as the steepness or gradient of a unit of terrain, usually measured as an angle in degrees or as a percentage. Aspect can be defined as the direction in which a unit of terrain faces. Aspect is usually expressed in degrees from north. Slope, aspect, and surface curvature in terrain analysis are all derived from neighborhood operations using elevation values of a cell's adjacent neighbours. Slope is a function of resolution, and the spatial resolution used to calculate slope and aspect should always be specified. Authors such as Skidmore, Jones and Zhou and Liu have compared techniques for calculating slope and aspect.

The following method can be used to derive slope and aspect:

The elevation at a point or unit of terrain will have perpendicular tangents (slope) passing through the point, in an east-west and north-south direction. These two tangents give two components, $\partial z/\partial x$ and $\partial z/\partial y$, which then be used to determine the overall direction of slope, and the aspect of the slope. The gradient is defined as a vector quantity with components equal to the partial derivatives of the surface in the x and y directions.

The calculation of the overall 3x3 grid slope S and aspect A for methods that determine east-west and north-south component use the following formulas respectively:

$$\tan S = \sqrt{\left(\frac{\partial z}{\partial x}\right)^2 + \left(\frac{\partial z}{\partial y}\right)^2}$$

$$\tan A = \left(\frac{\left(\frac{-\partial z}{\partial y}\right)}{\left(\frac{\partial z}{\partial x}\right)}\right)$$

Zhou and Liu describe another formula for calculating aspect, as follows:

$$A = 270^\circ + \arctan\left(\frac{\left(\frac{\partial z}{\partial x}\right)}{\left(\frac{\partial z}{\partial y}\right)}\right) - 90^\circ\left(\frac{\left(\frac{\partial z}{\partial y}\right)}{\left|\frac{\partial z}{\partial y}\right|}\right)$$

Data analysis

It is difficult to relate wetlands maps to rainfall amounts recorded at different points such as airports, television stations, and schools. A GIS, however, can be used to depict two- and three-dimensional characteristics of the Earth's surface, subsurface, and atmosphere from information points. For example, a GIS can quickly generate a map with isopleth or contour lines that indicate differing amounts of rainfall. Such a map can be thought of as a rainfall contour map. Many sophisticated methods can estimate the characteristics of surfaces from a limited number of point measurements. A two-dimensional contour map created from the surface modeling of rainfall point measurements may be overlaid and analyzed with any other map in a GIS covering the same area. This GIS derived map can then provide additional information - such as the viability of water power potential as a renewable energy source. Similarly, GIS can be used to compare other renewable energy resources to find the best geographic potential for a region.

Additionally, from a series of three-dimensional points, or digital elevation model, isopleth lines representing elevation contours can be generated, along with slope analysis, shaded relief, and other elevation products. Watersheds can be easily defined for any given reach, by computing all of the areas contiguous and uphill from any given point of interest. Similarly, an expected thalweg of where surface water would want to travel in intermittent and permanent streams can be computed from elevation data in the GIS.

Topological Modeling

A GIS can recognize and analyze the spatial relationships that exist within digitally stored spatial data. These topological relationships allow complex spatial modelling and analysis to be performed. Topological relationships between geometric entities traditionally include adjacency (what adjoins what), containment (what encloses what), and proximity (how close something is to something else).

Geometric Networks

Geometric networks are linear networks of objects that can be used to represent interconnected features, and to perform special spatial analysis on them. A geometric network is composed of edges, which are connected at junction points, similar to graphs in mathematics and computer science. Just like graphs, networks can have weight and flow assigned to its edges, which can be used to represent various interconnected features more accurately. Geometric networks are often used to model road networks and public utility networks, such as electric, gas, and water networks. Network modeling is also commonly employed in transportation planning, hydrology modeling, and infrastructure modeling.

Hydrological Modeling

GIS hydrological models can provide a spatial element that other hydrological models lack, with the analysis of variables such as slope, aspect and watershed or catchment area. Terrain analysis is fundamental to hydrology, since water always flows down a slope. As basic terrain analysis of a digital elevation model (DEM) involves calculation of slope and aspect, DEMs are very useful for hydrological analysis. Slope and aspect can then be used to determine direction of surface runoff, and hence flow accumulation for the formation of streams, rivers and lakes. Areas of divergent flow can also give a clear indication of the boundaries of a catchment. Once a flow direction and accumulation matrix has been created, queries can be performed that show contributing or dispersal areas at a certain point. More detail can be added to the model, such as terrain roughness, vegetation types and soil types, which can influence infiltration and evapotranspiration rates, and hence influencing surface flow. One of the main uses of hydrological modeling is in environmental contamination research.

Cartographic Modeling

The term “cartographic modeling” was probably coined by Dana Tomlin in his PhD dissertation and later in his book which has the term in the title. Cartographic modeling refers to a process where several thematic layers of the same area are produced, processed, and analyzed. Tomlin used raster layers, but the overlay method can be used more generally. Operations on map layers can be combined into algorithms, and eventually into simulation or optimization models.

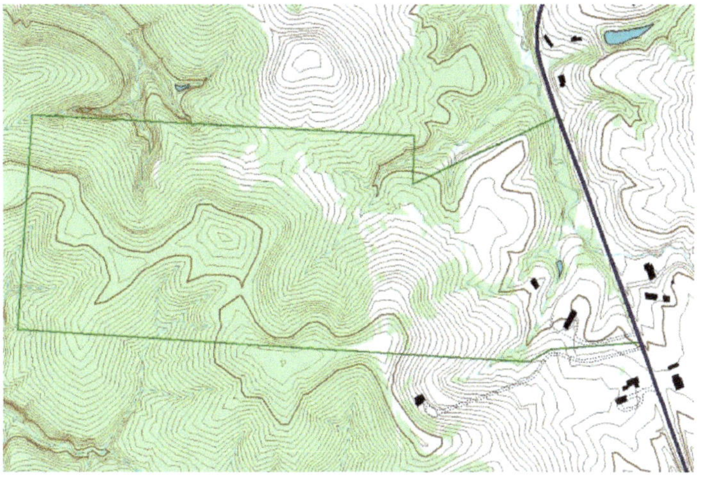

An example of use of layers in a GIS application. In this example, the forest cover layer (light green) is at the bottom, with the topographic layer over it. Next up is the stream layer, then the boundary layer, then the road layer. The order is very important in order to properly display the final result. Note that the pond layer was located just below the stream layer, so that a stream line can be seen overlying one of the ponds.

Map Overlay

The combination of several spatial datasets (points, lines, or polygons) creates a new output vector dataset, visually similar to stacking several maps of the same region. These overlays are similar to mathematical Venn diagram overlays. A union overlay combines the geographic features and attribute tables of both inputs into a single new output. An intersect overlay defines the area where both inputs overlap and retains a set of attribute fields for each. A symmetric difference overlay defines an output area that includes the total area of both inputs except for the overlapping area.

Data extraction is a GIS process similar to vector overlay, though it can be used in either vector or raster data analysis. Rather than combining the properties and features of both datasets, data extraction involves using a "clip" or "mask" to extract the features of one data set that fall within the spatial extent of another dataset.

In raster data analysis, the overlay of datasets is accomplished through a process known as "local operation on multiple rasters" or "map algebra," through a function that combines the values of each raster's matrix. This function may weigh some inputs more than others through use of an "index model" that reflects the influence of various factors upon a geographic phenomenon.

Geostatistics

Geostatistics is a branch of statistics that deals with field data, spatial data with a continuous index. It provides methods to model spatial correlation, and predict values at arbitrary locations (interpolation).

When phenomena are measured, the observation methods dictate the accuracy of any subsequent analysis. Due to the nature of the data (e.g. traffic patterns in an urban environment; weather patterns over the Pacific Ocean), a constant or dynamic degree of precision is always lost in the measurement. This loss of precision is determined from the scale and distribution of the data collection.

To determine the statistical relevance of the analysis, an average is determined so that points (gradients) outside of any immediate measurement can be included to determine their predicted behavior. This is due to the limitations of the applied statistic and data collection methods, and interpolation is required to predict the behavior of particles, points, and locations that are not directly measurable.

Hillshade model derived from a Digital Elevation Model of the Valestra area in the northern Apennines (Italy)

Interpolation is the process by which a surface is created, usually a raster dataset, through the input of data collected at a number of sample points. There are several forms of interpolation, each which treats the data differently, depending on the properties of the data set. In comparing interpolation methods, the first consideration should be whether or not the source data will change (exact or approximate). Next is whether the method is subjective, a human interpretation, or objective. Then there is the nature of transitions between points: are they abrupt or gradual. Finally, there is whether a method is global (it uses the entire data set to form the model), or local where an algorithm is repeated for a small section of terrain.

Interpolation is a justified measurement because of a spatial autocorrelation principle that recognizes that data collected at any position will have a great similarity to, or influence of those locations within its immediate vicinity.

Digital elevation models, triangulated irregular networks, edge-finding algorithms, Thiessen polygons, Fourier analysis, (weighted) moving averages, inverse distance weighting, kriging, spline, and trend surface analysis are all mathematical methods to produce interpolative data.

Address Geocoding

Geocoding is interpolating spatial locations (X,Y coordinates) from street addresses or any other spatially referenced data such as ZIP Codes, parcel lots and address locations. A reference theme is required to geocode individual addresses, such as a road centerline file with address ranges. The individual address locations have historically been interpolated, or estimated, by examining address ranges along a road segment. These are usually provided in the form of a table or database. The software will then place a dot approximately where that address belongs along the segment of centerline. For example, an address point of 500 will be at the midpoint of a line segment that starts with address 1 and ends with address 1,000. Geocoding can also be applied against actual parcel data, typically from municipal tax maps. In this case, the result of the geocoding will be an actually positioned space as opposed to an interpolated point. This approach is being increasingly used to provide more precise location information.

Reverse Geocoding

Reverse geocoding is the process of returning an estimated street address number as it relates to a given coordinate. For example, a user can click on a road centerline theme (thus providing a coordinate) and have information returned that reflects the estimated house number. This house number is interpolated from a range assigned to that road segment. If the user clicks at the midpoint of a segment that starts with address 1 and ends with 100, the returned value will be somewhere near 50. Note that reverse geocoding does not return actual addresses, only estimates of what should be there based on the predetermined range.

Multi-criteria Decision Analysis

Coupled with GIS, multi-criteria decision analysis methods support decision-makers in analysing a set of alternative spatial solutions, such as the most likely ecological habitat for restoration, against multiple criteria, such as vegetation cover or roads. MCDA uses decision rules to aggregate the criteria, which allows the alternative solutions to be ranked or prioritised. GIS MCDA may reduce costs and time involved in identifying potential restoration sites.

Data Output and Cartography

Cartography is the design and production of maps, or visual representations of spatial data. The vast majority of modern cartography is done with the help of computers, usually using GIS but production of quality cartography is also achieved by importing layers into a design program to refine it. Most GIS software gives the user substantial control over the appearance of the data.

Cartographic work serves two major functions:

First, it produces graphics on the screen or on paper that convey the results of analysis to the people who make decisions about resources. Wall maps and other graphics can be generated, allowing the viewer to visualize and thereby understand the results of analyses or simulations of potential events. Web Map Servers facilitate distribution of generated maps through web browsers using various implementations of web-based application programming interfaces (AJAX, Java, Flash, etc.).

Second, other database information can be generated for further analysis or use. An example would be a list of all addresses within one mile (1.6 km) of a toxic spill.

Graphic Display Techniques

Traditional maps are abstractions of the real world, a sampling of important elements portrayed on a sheet of paper with symbols to represent physical objects. People who use maps must interpret these symbols. Topographic maps show the shape of land surface with contour lines or with shaded relief.

Today, graphic display techniques such as shading based on altitude in a GIS can make relationships among map elements visible, heightening one's ability to extract and analyze information. For example, two types of data were combined in a GIS to produce a perspective view of a portion of San Mateo County, California.

- The digital elevation model, consisting of surface elevations recorded on a 30-meter horizontal grid, shows high elevations as white and low elevation as black.
- The accompanying Landsat Thematic Mapper image shows a false-color infrared image looking down at the same area in 30-meter pixels, or picture elements, for the same coordinate points, pixel by pixel, as the elevation information.

A GIS was used to register and combine the two images to render the three-dimensional perspective view looking down the San Andreas Fault, using the Thematic Mapper image pixels, but shaded using the elevation of the landforms. The GIS display depends on the viewing point of the observer and time of day of the display, to properly render the shadows created by the sun's rays at that latitude, longitude, and time of day.

An archeochrome is a new way of displaying spatial data. It is a thematic on a 3D map that is applied to a specific building or a part of a building. It is suited to the visual display of heat-loss data.

Spatial ETL

Spatial ETL tools provide the data processing functionality of traditional Extract, Transform, Load (ETL) software, but with a primary focus on the ability to manage spatial data. They provide GIS users with the ability to translate data between different standards and proprietary formats, whilst geometrically transforming the data en route. These tools can come in the form of add-ins to existing wider-purpose software such as Microsoft Excel.

GIS Data Mining

GIS or spatial data mining is the application of data mining methods to spatial data. Data mining, which is the partially automated search for hidden patterns in large databases, offers great potential benefits for applied GIS-based decision making. Typical applications include environmental monitoring. A characteristic of such applications is that spatial correlation between data measurements require the use of specialized algorithms for more efficient data analysis.

Applications

The implementation of a GIS is often driven by jurisdictional (such as a city), purpose, or application requirements. Generally, a GIS implementation may be custom-designed for an organization. Hence, a GIS deployment developed for an application, jurisdiction, enterprise, or purpose may not be necessarily interoperable or compatible with a GIS that has been developed for some other application, jurisdiction, enterprise, or purpose.

GIS provides, for every kind of location-based organization, a platform to update geographical data without wasting time to visit the field and update a database manually. GIS when integrated with other powerful enterprise solutions like SAP and the Wolfram Language helps creating powerful decision support system at enterprise level.

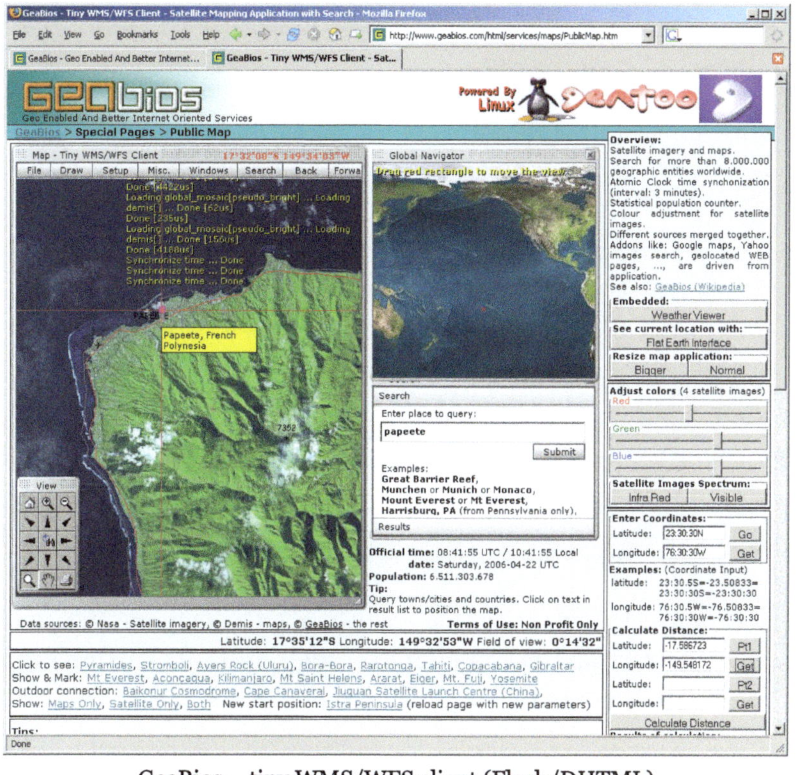

GeaBios – tiny WMS/WFS client (Flash/DHTML)

Many disciplines can benefit from GIS technology. An active GIS market has resulted in lower costs and continual improvements in the hardware and software components of GIS, and usage in the fields of science, government, business, and industry, with applications including real estate, public health, crime mapping, national defense, sustainable development, natural resources, climatology, landscape architecture, archaeology, regional and community planning, transportation and logistics. GIS is also diverging into location-based services, which allows GPS-enabled mobile devices to display their location in relation to fixed objects (nearest restaurant, gas station, fire hydrant) or mobile objects (friends, children, police car), or to relay their position back to a central server for display or other processing.

Open Geospatial Consortium Standards

The Open Geospatial Consortium (OGC) is an international industry consortium of 384 companies, government agencies, universities, and individuals participating in a consensus process to develop publicly available geoprocessing specifications. Open interfaces and protocols defined by OpenGIS Specifications support interoperable solutions that "geo-enable" the Web, wireless and location-based services, and mainstream IT, and empower technology developers to make complex spatial information and services accessible and useful with all kinds of applications. Open Geospatial Consortium protocols include Web Map Service, and Web Feature Service.

GIS products are broken down by the OGC into two categories, based on how completely and accurately the software follows the OGC specifications.

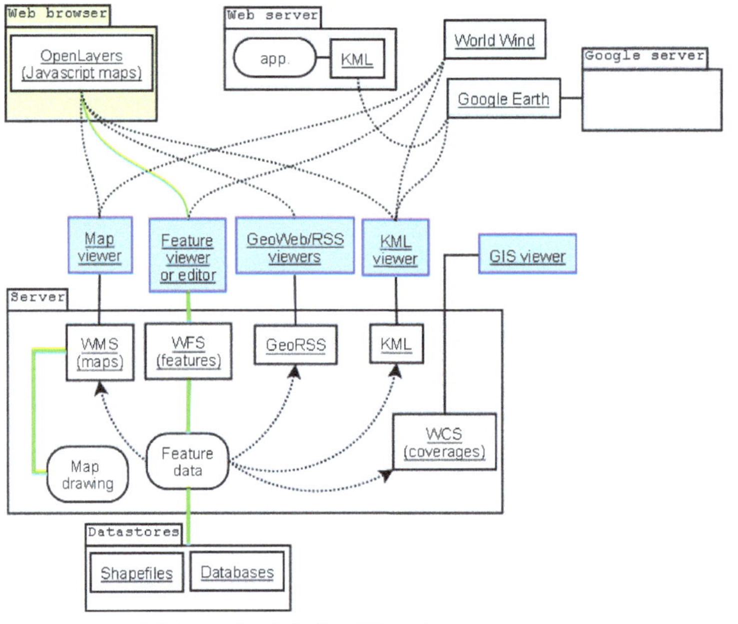

OGC standards help GIS tools communicate.

Compliant Products are software products that comply to OGC's OpenGIS Specifications. When a product has been tested and certified as compliant through the OGC Testing Program, the product is automatically registered as "compliant" on this site.

Implementing Products are software products that implement OpenGIS Specifications but have not yet passed a compliance test. Compliance tests are not available for all specifications. Developers can register their products as implementing draft or approved specifications, though OGC reserves the right to review and verify each entry.

Web Mapping

In recent years there has been an explosion of mapping applications on the web such as Google Maps and Bing Maps. These websites give the public access to huge amounts of geographic data.

Some of them, like Google Maps and OpenLayers, expose an API that enable users to create custom applications. These toolkits commonly offer street maps, aerial/satellite imagery, geocoding, searches, and routing functionality. Web mapping has also uncovered the potential of crowdsourcing geodata in projects like OpenStreetMap, which is a collaborative project to create a free editable map of the world.

Adding the Dimension of Time

The condition of the Earth's surface, atmosphere, and subsurface can be examined by feeding satellite data into a GIS. GIS technology gives researchers the ability to examine the variations in Earth processes over days, months, and years. As an example, the changes in vegetation vigor through a growing season can be animated to determine when drought was most extensive in a particular region. The resulting graphic represents a rough measure of plant health. Working with two variables over time would then allow researchers to detect regional differences in the lag between a decline in rainfall and its effect on vegetation.

GIS technology and the availability of digital data on regional and global scales enable such analyses. The satellite sensor output used to generate a vegetation graphic is produced for example by the Advanced Very High Resolution Radiometer (AVHRR). This sensor system detects the amounts of energy reflected from the Earth's surface across various bands of the spectrum for surface areas of about 1 square kilometer. The satellite sensor produces images of a particular location on the Earth twice a day. AVHRR and more recently the Moderate-Resolution Imaging Spectroradiometer (MODIS) are only two of many sensor systems used for Earth surface analysis. More sensors will follow, generating ever greater amounts of data.

In addition to the integration of time in environmental studies, GIS is also being explored for its ability to track and model the progress of humans throughout their daily routines. A concrete example of progress in this area is the recent release of time-specific population data by the U.S. Census. In this data set, the populations of cities are shown for daytime and evening hours highlighting the pattern of concentration and dispersion generated by North American commuting patterns. The manipulation and generation of data required to produce this data would not have been possible without GIS.

Using models to project the data held by a GIS forward in time have enabled planners to test policy decisions using spatial decision support systems.

Semantics

Tools and technologies emerging from the W3C's Data Activity are proving useful for data integration problems in information systems. Correspondingly, such technologies have been proposed as a means to facilitate interoperability and data reuse among GIS applications. and also to enable new analysis mechanisms.

Ontologies are a key component of this semantic approach as they allow a formal, machine-readable specification of the concepts and relationships in a given domain. This in turn allows a GIS to focus on the intended meaning of data rather than its syntax or structure. For example, reasoning that a land cover type classified as *deciduous needleleaf trees* in one dataset is a specialization or subset of land cover type *forest* in another more roughly classified dataset can help a GIS automatically merge the two datasets under the more general land cover classification. Tentative ontologies have been developed in areas related to GIS applications, for example the hydrology ontology developed by the Ordnance Survey in the United Kingdom and the SWEET ontologies developed by NASA's Jet Propulsion Laboratory. Also, simpler ontologies and semantic metadata standards are being proposed by the W3C Geo Incubator Group to represent geospatial data on the web. GeoSPARQL is a standard developed by the Ordnance Survey, United States Geological Survey, Natural Resources Canada, Australia's Commonwealth Scientific and Industrial Research Organisation and others to support ontology creation and reasoning using well-understood OGC literals (GML, WKT), topological relationships (Simple Features, RCC8, DE-9IM), RDF and the SPARQL database query protocols.

Recent research results in this area can be seen in the International Conference on Geospatial Semantics and the Terra Cognita – Directions to the Geospatial Semantic Web workshop at the International Semantic Web Conference.

Implications of GIS in Society

With the popularization of GIS in decision making, scholars have begun to scrutinize the social and political implications of GIS. GIS can also be misused to distort reality for individual and political gain. It has been argued that the production, distribution, utilization, and representation of geographic information are largely related with the social context and has the potential to increase citizen trust in government. Other related topics include discussion on copyright, privacy, and censorship. A more optimistic social approach to GIS adoption is to use it as a tool for public participation.

i-Tree

i-Tree is a collection of software programs for urban forestry that anyone can use. It was designed and developed by the USDA Forest Service to place a value on urban forest ecosystem services such as pollution removal, carbon storage, avoided carbon emissions, and avoided storm water runoff. I-Tree provides baseline data so that the growth of trees can be followed over time, and is used for planning purposes. Reports are based on tree inventories of species, DBH (diameter at breast height) and tree height. I-Tree is peer-reviewed and has a process of ongoing collaboration to improve it.

History

I-Tree began in 2002 as survey of a sample of urban forest to simulate taking a tree inventory of an entire urban forest. It then added hand held devices for efficient inventory of street trees. The current version of i-Tree allows for several sources of data to be used, such as National Land Cover Data, Google Maps, and tree inventories. Some versions allow for continuous data on air pollution and meteorology in order to average the results for a more accurate answer.

The current version uses groundwater capture, building energy savings, air pollution benefits, carbon sequestration, homeowner value, and pest detection.

References

- "Geographic Information Systems as an Integrating Technology: Context, Concepts, and Definitions". Kenneth E. Foote and Margaret Lynch, The Geographer's Craft Project, Department of Geography, The University of Colorado at Boulder. Retrieved 21 Apr 2015.
- "Rapport sur la marche et les effets du choléra dans Paris et le département de la Seine. Année 1832". Gallica. Retrieved 10 May 2012.
- "Aeryon Announces Version 5 of the Aeryon Scout System | Aeryon Labs Inc". Aeryon.com. 2011-07-06. Retrieved 2012-05-13.

7

Measures of Product Standardization

In order to ensure legal and correct methods of tree-felling and timber production, organizations have created measures of standardisation that reflect the quality of the product as well as safe environmental conditions. Some measures include certified wood, old-growth forests and environmental audits. This chapter is a compilation of the various branches of forestry that form an integral part of the broader subject matter.

Certified Wood

Certified wood and paper products come from responsibly managed forests – as defined by a particular standard. With third-party forest certification, an independent organization develops standards of good forest management, and independent auditors issue certificates to forest operations that comply with those standards.

A managed Forest on San Juan Island in Washington (U.S. state).

Requirements

Forest certification programs typically require that forest management practices conform to existing laws. Other basic requirements or characteristics of forest certification programs include:

Basic requirements of credible forest certification programs include:

- Protection of biodiversity, species at risk and wildlife habitat; sustainable harvest levels; protection of water quality; and prompt regeneration (e.g., replanting and reforestation).
- Third-party certification audits performed by accredited certification bodies.

- Publicly available certification audit summaries.
- Multi-stakeholder involvement in a standards development process.
- Complaints and appeals process.

Programs

Today there are more than 50 certification programs worldwide addressing the many types of forests and tenures around the world. The two largest international forest certification programs are the Forest Stewardship Council (FSC) and the Programme for the Endorsement of Forest Certification (PEFC).

The PEFC is the largest certification framework in terms of forest area, with approximately two-thirds of the total certified area. The FSC program is the fastest growing.

Third-party forest certification was pioneered in the early 1990s by the FSC, a collaboration between environmental NGOs, forest product companies and social interests. Competing systems quickly emerged throughout the world. Some commentators, including Jared Diamond, have suggested that many competing standards were set up by logging companies specifically aiming to confuse consumers with less rigorously enforced but similarly named competing standards.

United States and Canada

In the United States and Canada, there are a number of forest certification programs. Three of these programs are endorsed by the PEFC. They are the American Tree Farm System (ATFS), the Canadian Standards Association's Sustainable Forest Management Standard and the Sustainable Forestry Initiative (SFI) Program. ATFS is applicable only in the United States; the Canadian Standards Association SFM Standard is applicable only in Canada. SFI is applicable to both the United States and Canada. The FSC, program is applied throughout North America. SFI is the world's largest regional forest certification program in terms of total certified forest area.

The National Association of State Foresters in the USA passed a resolution in 2008 that supports *all* of the forest certification systems used in the USA and recognized the value of their differences: "... the ATFS, FSC, and SFI systems include the fundamental elements of credibility and make positive contributions to forest sustainability.... No certification program can credibly claim to be 'best', and no certification program that promotes itself as the only certification option can maintain credibility. Forest ecosystems are complex and a simplistic 'one size fits all' approach to certification cannot address all sustainability needs.".

The Canadian Council of Forest Ministers issued a statement in 2008 on forest certification standards in Canada, which said: "In Canada, each jurisdiction's forest laws, policies and administrative requirements comprise an over-arching framework that fully characterizes what sustainable forest management (SFM) means in that jurisdiction, and what actions may take place on public and/or private forest land. Governments in Canada support third-party forest certification as a tool to demonstrate the rigor of Canada's forest management laws, and to document the country's world-class sustainable forest management record. The forest management standards of the Canadian Standards Association (CSA), the FSC and the Sustainable Forestry Initiative (SFI) are all used in Canada. Governments in Canada accept that these standards demonstrate, and promote the sustainability of forest management practices in Canada."

Chain-of-custody Certification

Chain of Custody certification tracks the certified material through the production process – from the forest to the consumer, including all successive stages of processing, transformation, manufacturing and distribution. It provides evidence that certified material in a certified product originates from certified forests.

The United Nations reports that between January 2009 and May 2010, the total number of PEFC and FSC chain-of-custody certificates issued worldwide increased by 88% for a total of 23,717 certificates (this does not include SFI certificates).

Future Expansion

Forest certification is a voluntary process about 10% of the world's forest under at least one certification program. Customers that choose to buy certified products are supporting land managers, land owners and forest product companies that have made a commitment to meeting the standards of forest certification.

Third-party forest certification is a useful tool for those seeking to purchase paper and wood products that come from forests that are well-managed and use materials that are legally harvested. Incorporating third-party certification into forest product buying practices can be a centerpiece for responsible wood and paper purchasing policies that include factors such as the protection of sensitive forest values, thoughtful material selection and efficient use of products.

The 2009-2010 United Nations Market Review reported that companies that produced or traded in certified forest products often had a market advantage during the 2008-2009 recession because, in a buyers' market, buyers could be more selective in choosing their sources of supply. The report cites four demand drivers for certification:

- Paper, publishing, printing and packaging – commitments to increase the use of responsible paper sources by large publishers such as Time Inc. has probably been the most significant factor driving growth in forest and chain-of-custody certification.
- Green public procurement – governments such as the UK and the Netherlands have adopted green timber procurement policies, including recognition of FSC and PEFC endorsed programs. An example is the UK's Central Point of Expertise on Timber (set up by the Department of Environment, Food and Rural Affairs and operated by ProForest).
- Green building – standards for green building incentivize and reward the use of certified wood products.
- Illegal logging – new legislation designed to minimize the risk of illegal wood entering supply chains such as the amended Lacey Act in the United States has created a strong incentive to demand independently certified wood that can address illegal logging concerns.

The World Resources Institute, in partnership with the Environmental Investigation Agency, released a fact sheet designed to answer some of the frequently asked questions about the Lacey Act, which was amended in 2008 to ban commerce in illegally sourced plants and their products—including timber, wood, and paper products. The fact sheet says forest certification is a very good approach for demonstrating due care by showing government and customers that a company has taken proactive steps to eliminate illegal wood or plant material from its supply chain. Certification does not relieve importers of the requirement to submit appropriate import declaration information to U.S. government agencies.

CarbonFix Standard

The CarbonFix Standard (CFS) is an initiative supported by over 60 organisations promotes the development of climate forestation projects to sequester carbon from the atmosphere. The CarbonFix Standard is administered by CarbonFix, a non-profit association based in Germany.

The CarbonFix Standard has been favourably assessed in 2008, by the University of Canterbury, New Zealand and in 2009 by the Imperial College of London. CarbonPositive (2009) placed the CarbonFix Standard between the Voluntary Carbon Standard and the CCB Standard, "aiming for a best of both worlds approach".

History

The association CarbonFix was founded in 1999. In 2007, the association developed the first version of the CarbonFix Standard, which was presented at the World Climate Conference in Bali, in December 2007. Since then, the standard has been revised several times.

Structure

The *Criteria and Methodology* section lists the criteria a project has to meet to become certified. Tree planting activities must take place on land which contained no forest before. Only projects which require the additional financial means from the sale of CO_2 credits will be certified. There must be Sustainable Forest Management. CO2-Fixation is verified using methodology based on the rules of the United Nations Climate Secretariat.

The *Procedures* of the CarbonFix Standard describes how to prepare project information and request certification, how to sell CO_2-certificates, what fees CarbonFix charges and how CO_2-certificates are placed into a carbon registry.

Programme for the Endorsement of Forest Certification

The Programme for the Endorsement of Forest Certification (PEFC) is an international, non-profit, non-governmental organization which promotes sustainable forest management through independent third party certification. It is considered the certification system of choice for small forest owners. LUL

Its 35 worldwide independent national forest certification systems represent more than 240 million ha of certified forests, making it the largest forest certification system in the world, covering about two-thirds of the globally certified forest area. It is based in Geneva, Switzerland.

History

PEFC was founded in 1999 in response to the specific requirements of small- and family forest owners as an international umbrella organization providing independent assessment, endorsement and recognition of national forest certification systems. It responded to the need for a mechanism enabling the independent development of national standards tailored to the political, economic, social, environmental and cultural realities of the respective countries, while at the same time ensuring compliance with internationally accepted requirements and global recognition.

After the successful endorsement of certification systems in Europe, Australia and Chile became the first non-European national standards to be endorsed by PEFC in 2004. PEFC's certification criteria are based on globally recognized principles, guidelines and criteria developed by international and intergovernmental bodies with broad consensus from interested stakeholders.

Today, PEFC is the world's largest forest certification system and the certification system of choice for small forest owners.

Sustainable Forest Management Criteria

Shield of the Programme for the Endorsement of Forest Certification Schemes (PEFC) at Marburg-Schröck, Germany.

PEFC International is the only international forest certification scheme that bases its criteria on internationally accepted intergovernmental conventions and guidelines, thereby linking its sustainability benchmark criteria with existing governmental processes. This includes:

- Pan-European Criteria, Indicators and Operational Level Guidelines for Sustainable Forest Management (Ministerial Conference on the Protection of Forests in Europe)
- ATO/ITTO Principles, criteria and indicators for the sustainable forest management of African natural tropical forests (ATO/ITTO)
- ITTO guidelines on sustainable forest management (ITTO)

PEFC requires adherence to all eight core ILO conventions, even in countries which have not ratified them. These conventions are

- No 29: Forced Labour (1930)
- No 87: Freedom of Associations and Protection of the Right to Organise (1948)
- No 98: Right to Organise and Collective Bargaining (1949)
- No 100: Equal Remuneration (1951)
- No 105: Abolition of Forced Labour (1957)
- No 111: Discrimination (Employment and Occupation) (1958)
- No 138: Minimum Age for Admission to Employment (1973)
- No. 182: Worst Forms of Child Labour (1999)

National Forest Certification Systems

PEFC only recognize forests certified to standards that have been reviewed and endorsed by PEFC.

National forest certification systems that wish to be recognized by are required to set standards keeping with the requirements of ISO/IEC Guide 59:1994 Code of good practice for standardization. National standard must be developed by so-called National Governing Bodies, and meet requirements for transparency, consultation and decision-making by consensus. These guidelines also outline processes for revising and amending standards, and provide those who utilise the standard with the security of future certainty.

Endorsements

All PEFC-endorsed standards have been subjected to rigorous public review during their development. National forest certification systems wanting to obtain PEFC endorsement are subject to an independent assessment to ensure that it meets the many PEFC requirements for the standards development process, public review and forest management requirements. The consultant's report is reviewed by an independent Panel of Experts and the PEFC Board, and if satisfactory, the new standard is approved by the PEFC members as a PEFC-endorsed standard.

In line with its commitment to transparency, PEFC makes its entire documentation of national forest certification system, including the independent assessments, publicly available.

Countries

Countries with PEFC endorsed national certification systems include: Argentina, Australia, Austria, Belarus, Belgium, Brazil, Canada, Chile, China, Czech Republic, Denmark, Estonia, Finland, France, Gabon, Germany, Indonesia, Ireland, Italy, Latvia, Luxembourg, Malaysia, Netherlands, Norway, Poland, Portugal, Russia, Slovak Republic, Slovenia, Spain, Sweden, Switzerland, United Kingdom, United States, and Uruguay.

Criticism and Alternative Certification Schemes

Forest Stewardship Council is the main alternative forest certification system. Mutual recognition of FSC and PEFC certified material in the chain of custody has not yet happened. However, FSC and PEFC use the same forest management standard in countries such as the United Kingdom, Switzerland and Norway; Malaysia has submitted its timber certification scheme for endorsement by PEFC that is largely based on FSC principles and criteria as the template.

Several environmental non-governmental organisations, such as The Wilderness Society, Greenpeace and FERN have criticised the PEFC. Greenpeace does not believe alternatives to the FSC, including PEFC, can ensure responsible forest management.

The PEFC was introduced as a direct alternative to the more environmentally created schemes in Germany.

Old-growth Forest

Old-growth European Beech forest in Biogradska Gora National Park, Montenegro

An old-growth forest — also termed primary forest, virgin forest, primeval forest, late seral forest, or (in Britain) ancient woodland — is a forest that has attained great age without significant disturbance and thereby exhibits unique ecological features and might be classified as a climax community. Old-growth features include diverse tree-related structures that provide diverse wildlife habitat that increases the bio-diversity of the forested ecosystem. The concept of diverse tree structure includes multi-layered canopies and canopy gaps, greatly varying tree heights and diameters, and diverse tree species and classes and sizes of woody debris.

Old-growth forests are economically valuable, and logging of these forests has been a point of contention between the logging industry and environmentalists.

Coast Redwoods in old-growth forest in Muir Woods National Monument, Marin County, California

Characteristics

Tongass National Forest, Alaska

Old-growth forests tend to have large trees and standing dead trees, multi-layered canopies with gaps that result from the deaths of individual trees, and coarse woody debris on the forest floor.

Forest regenerated after a severe disturbance, such as wildfire, insect infestations or harvesting, is often called second-growth or *regeneration* until enough time passes for the effects of the disturbance to be no longer evident. Depending on the forest, this may take anywhere from a century to several millennia. Hardwood forests of the eastern United States can develop old-growth characteristics in one or two generations of trees, or 150–500 years. In British Columbia, Canada, old growth is defined as 120 to 140 years of age in the interior of the province where fire is a frequent and natural occurrence. In British Columbia's coastal rainforests, old growth is defined as trees more than 250 years, with some trees reaching more than 1,000 years of age. In Australia, eucalypt trees rarely exceed 350 years of age due to frequent fire disturbance.

Forest types have very different development patterns, natural disturbances and appearances. A Douglas-fir stand may grow for centuries without disturbance while an old-growth ponderosa pine forest requires frequent surface fires to reduce the shade-tolerant species and regenerate the canopy species. In the Boreal-West Forest Region, catastrophic disturbances like wildfires minimize opportunities for major accumulations of dead and downed woody material and other structural legacies associated with old growth conditions. Typical characteristics of old-growth forest include presence of older trees, minimal signs of human disturbance, mixed-age stands, presence of canopy openings due to tree falls, pit-and-mound topography, down wood in various stages of decay, standing snags (dead trees), multi-layered canopies, intact soils, a healthy fungal ecosystem, and presence of indicator species.

Biodiversity

The northern spotted owl primarily inhabits old growth forests in the northern part of its range (Canada to southern Oregon) and landscapes with a mix of old and younger forest types in the southern part of its range (Klamath region and California).

Old-growth forests are often biologically diverse, and home to many rare species, threatened species, and endangered species of plants and animals, such as the northern spotted owl, marbled murrelet and fisher, making them ecologically significant. Levels of biodiversity may be higher or lower in old-growth forests compared to that in second-growth forests, depending on specific circumstances, environmental variables and geographic variables. Logging in old-growth forests is a contentious issue in many parts of the world. Excessive logging reduces biodiversity, affecting not only the old-growth forest itself, but also indigenous species that rely upon old-growth forest habitat.

Mixed Age

A forest in old-growth stage has a mix of tree ages, due to a distinct regeneration pattern for this stage. New trees regenerate at different times from each other, because each one of them has different spatial location relative to the main canopy and hence each one receives a different amount of light. The mixed age of the forest is an important criterion in ensuring that the forest is a relatively stable ecosystem in the long term. A climax stand that is uniformly aged becomes senescent and degrades within a relatively short time-period to result in a new cycle of forest succession. Thus, uniformly aged stands are a less stable ecosystem.

Canopy Openings

Forest canopy gaps are essential in creating and maintaining mixed-age stands. Also, some herbaceous plants only become established in canopy openings, but persist beneath an understory. Openings are a result of tree death due to small impact disturbances such as wind, low-intensity fires and tree diseases.

Old-growth forests are unique, usually having multiple horizontal layers of vegetation representing a variety of tree species, age classes, and sizes, as well as "pit and mound" soil shape with well-established fungal nets. Because old-growth forest is structurally diverse it provides higher-diversity habitat than forests in other stages. Thus, sometimes higher biological diversity can be sustained in old-growth forest, or at least a biodiversity that is different from other forest stages.

Virgin forest at approx 2500m above sea level in Shennongjia Forestry District, Hubei, China

Topography

The characteristic topography of much old-growth forest consists of pits and mounds. Mounds are caused by decaying fallen trees, and pits (tree throws) by the roots pulled out of the ground when trees fall due to natural causes, including being pushed over by animals. Pits expose humus-poor, mineral-rich soil and often collect moisture and fallen leaves, forming a thick organic layer that is able to nurture certain types of organisms. Mounds provide a place free of leaf inundation and saturation, where other types of organisms thrive.

Standing Snags

Standing snags provide food sources and habitat for many types of organisms. In particular, many species of dead-wood predators such as woodpeckers must have standing snags available for feeding. In North America the spotted owl is well known for needing standing snags for nesting habitat.

Decaying Ground Layer

Downed wood replenishes topsoil as it decays

Fungus on a tree stump in the Białowieża Forest, one of the last largely intact primeval forests in Central Europe

Fallen timber, or coarse woody debris, contributes carbon-rich organic matter directly to the soil, providing a substrate for mosses, fungi, and seedlings, and creating microhabitats by creating relief on the forest floor. In some ecosystems such as the temperate rain forest of the North American Pacific coast, fallen timber may become nurse logs, providing a substrate for seedling trees.

Soil

Intact soils harbor many life-forms that rely on them. Intact soils generally have very well-defined horizons, or soil profiles. Different organisms may need certain well-defined soil horizons in order to live, while many trees need well-structured soils free of disturbance in order to thrive. Some herbaceous plants in northern hardwood forests must have thick duff layers (which are part of the soil profile). Fungal ecosystems are essential for efficient *in-situ* recycling of nutrients back into the entire ecosystem.

Ecological Definitions

Stand Age Definition

Stand age can also be used to categorize forest as old-growth. For each geographical area, there is an average time since disturbance when the forest will reach old-growth stage. This method is useful, because it allows quick and objective determination of forest stage. However, this definition does not provide explanation about forest function. It just gives a useful number to measure. Due to that fact, some forests may be excluded from being categorized as old-growth even if they have old-growth attributes just because they are too young. Also, older forests can lack some old-growth attributes and be categorized as old-growth just because they are so old. The idea of using age is also problematic, because human activities can influence the forest in varied ways. For example, after logging of 30% of the trees, we can wait less time for old-growth to come back than after removal of 80% of the trees. Although depending on the species logged, the forest that comes back after a 30% harvest may consist of proportionately less hardwood trees than a forest logged at 80% in which the light competition by less important tree species does not inhibit the regrowth of vital hardwoods.

Forest Dynamics Definition

From a forest dynamics perspective, old-growth forest is a forest in a stage that follows Understory Reinitiation stage. A review of the stages helps to understand the concept:

1. Stand-replacing: disturbance hits the forest and kills most of the living trees.
2. Stand-initiation: population of new trees becomes established.
3. Stem-exclusion: trees grow higher and enlarge their canopy, thus competing for the light with neighbors. Light competition mortality kills slowly growing trees and reduces forest density. This allows surviving trees to increase in size. Eventually the canopies of neighboring trees touch each other and drastically lowers amount of light that reaches lower layers. Due to that, the understory dies and only very shade-tolerant species survive.
4. Understory reinitiation: trees die from low level mortality, such as windthrow and diseas-

es. Individual canopy gaps start to appear and more light can reach forest floor. Hence, shade-tolerant species can establish in the understory.

5. Old-growth: Main canopy trees become older and more of them die, creating even more gaps. Since the gaps appear at different times, the understory trees will be at different growth stages. Furthermore, the amount of light that reaches each understory tree depends on its position relative to the gap. Thus, each understory tree grows at a different speed. The difference in establishment timing and in growth speed create a population of understory trees that are variable in size. Eventually, some understory trees grow to become as tall as the main canopy trees, thereby filling the gap. This perpetuation process is typical for the old-growth stage. This, however, does not mean that the forest will be old-growth forever. Generally there are three possible futures for old-growth stage forest: (A) The forest will be hit by a new stand-replacing disturbance and most of the trees will die. (B) The tree community will eventually create unfavorable conditions for new trees to regenerate. In this case, the old trees will die and smaller plants will create woodland. (C) The regenerating understory trees are different species than the main canopy trees. In this case, the forest will switch back to Stem-Exclusion stage, but with different tree species. The forest in old-growth stage can be stable for centuries but the length of this stage depends on the forest's tree composition and climate of the area. For example, frequent natural fires do not allow boreal forests to be as old as coastal forests of western North America.

Of importance is that while the stand switches from one tree community to another, the stand will not necessarily go through old-growth stage between those stages. Some tree species have relatively open canopy. That allows more shade-tolerant tree species to establish below even before Understory Reinitiation stage. The shade-tolerant trees will eventually out-compete the main canopy trees in stem-exclusion stage. Therefore, the dominant tree species will change, but the forest will still be in Stem-Exclusion stage.

Tree species succession may change tree species composition once the old-growth stage has been achieved. For example, an old boreal forest may contain some large aspen trees, which may die and be replaced by smaller balsam fir or black spruce. Consequently, the forest will switch back to Understory Reinitiation stage. If old growth stage is seen as an end point of stand development, it can be easily evaluated using structural or static attributes. However, in some forest ecosystems this can lead to decisions regarding the preservation of unique stands or attributes that will disappear over the next few decades because of natural succession processes. Consequently, using stand dynamics to define old-growth forest is more useful in cases where the species that constitute old-growth forest can have long life span or in ecosystem where succession is very slow.

Social and Cultural Definitions

Common cultural definitions and common denominators regarding what comprises old-growth forest, and of the variables that define, constitute and embody old-growth forests include:

- The forest habitat possesses relatively mature, old trees;
- The old-growth trees have long continuity on the same site;

- The forest itself has not been subjected to significant inhabitation by mankind that has altered the appearance of the landscape and its ecosystems, has not been subjected to logging, and has inherently progressed per natural tendencies.

Redwood tree in northern California redwood forest. According to the National Park Service "96 percent of the original old-growth coast redwoods have been logged."

The debate over old growth definitions has been inextricably linked with a complex range of social perceptions about wilderness preservation, aesthetics and spirituality, as well as economic or industrial values.

Economic Definitions

Old-growth forests were often given harvesting priority because they have the most commercially valuable timber, they are considered to be at greater risk of deterioration through root rot or insect infestation, and they occupy land that could be used for more productive second-growth stands. In some regions, old growth is not the most commercially viable timber – in British Columbia, Canada, harvesting in the coastal region is moving to younger second-growth stands.

Other Definitions

A 2001 scientific symposium in Canada found that defining old growth in a scientifically meaningful, yet policy-relevant, manner presents some basic difficulties, especially if a simple, unambiguous, and rigorous scientific definition is sought. Symposium participants identified some attributes of late-successional, temperate-zone, old-growth forest types that could be considered in developing an index of "old-growthness" and for defining old-growth forests:

Structural Features:

- Uneven or multi-aged stand structure, or several identifiable age cohorts
- Average age of dominant species approaching half the maximum longevity for species (approximately 150+ years for most shade-tolerant trees)

- Some old trees at close to their maximum longevity (ages of 300+ years)
- Presence of standing dead and dying trees in various stages of decay
- Fallen, coarse woody debris
- Natural regeneration of dominant tree species within canopy gaps or on decaying logs

Avatar Grove near Port Renfrew, British Columbia. Giant Douglas-firs (left) and red cedars (right) fill the grove.

Compositional Features:

- Long-lived, shade-tolerant tree species associations (e.g., sugar maple, American beech, yellow birch, red spruce, eastern hemlock, white pine)

Process Features:

- Characterized by small-scale disturbances creating gaps in forest canopy
- A long natural rotation for catastrophic or stand-replacing disturbance (e.g., a period greater than the maximum longevity of the dominant tree species)
- Minimal evidence of human disturbance
- Final stages of stand development before a relatively steady state is reached

Importance

- Old-growth forests often contain rich communities of plants and animals within the habitat due to the long period of forest stability. These varied and sometimes rare species may depend on the unique environmental conditions created by these forests.
- Old-growth forest serves as a reservoir for species which cannot thrive or easily regenerate in younger forest, and so can be used as a baseline for research.
- Plant species that are native to old-growth forests may someday prove to be invaluable to-

wards curing various human ailments, as has been realized in numerous plants in tropical rainforests.

- Old-growth forests also store large amounts of carbon above and below the ground (either as humus, or in wet soils as peat). They collectively represent a very significant store of carbon. Destruction of these forests releases this carbon as greenhouse gases, and may increase the risk of global climate change.

Ecosystem Services

Old-growth forests provide ecosystem services that may be far more important to society than their use as a source of raw materials. These services include breathable air, pure water, carbon storage, regeneration of nutrients, maintenance of soils, pest control by insectivorous bats and insects, micro- and macro-climate control, and the storage of a wide variety of genes.

Climatic Impacts

The effects of old-growth forests in relation to Global Warming has been contested in various studies and journals.

The Intergovernmental Panel on Climate Change said in its 2007 report: "In the long term, a sustainable forest management strategy aimed at maintaining or increasing forest carbon stocks, while producing an annual sustained yield of timber, fibre or energy from the forest, will generate the largest sustained mitigation benefit."

Critics note that at old-growth forests are often perceived to be in equilibrium, but could be releasing as much carbon dioxide as they capture, or are currently in a state of decay. Another scientific study concluded that forest harvesting has little or no effect on the amount of carbon stored in the soil. As trees grow, they remove carbon from the atmosphere. As they reach maturity, growth slows and ultimately stops as mortality catches up to growth. Harvesting also removes carbon from the forest but some of it is stored in wood products (preventing its immediate release to the atmosphere) and some is available for use as biomass energy (displacing fossil fuel use), although using biomass as a fuel produces air pollution in the form of carbon monoxide, NOx (nitrogen oxides), VOCs (volatile organic compounds), particulates and other pollutants, in some cases at levels above those from traditional fuel sources such as coal or natural gas. In most North American forests, this drop happens when a tree is between 60 and 150 years old, depending on the species and environmental factors.

Each forest has a different potential to store carbon. For example, this potential is particularly high in the Pacific Northwest where forests are relatively productive, trees live a long time, decomposition is relatively slow, and fires are infrequent. The differences between forests must therefore be taken into consideration when determining how they should be managed to store carbon.

Logging in Old Growth Forests

According to the World Resources Institute, as of January 2009, only 21% of the original old-growth forests that once existed on earth are remaining. It is estimated that one half of Western Europe's forests were cleared before the Middle Ages, and that 90% of the old-growth forests that existed in the contiguous United States in the 1600s have been cleared.

Old growth red cedar stump near Port Renfrew, British Columbia

The large trees in old growth forests are economically valuable, and have been subjected to aggressive logging around the world. This has led to much controversy between logging companies and environmental groups. From certain forestry perspectives, fully maintaining an old growth forest is seen as extremely economically unproductive, as timber can only be collected from falling trees, and also potentially damaging to nearby managed groves by creating environments conducive to root rot. From this view, it may be more productive to cut the old growth down and replace the forest with a younger one. On the other hand, old growth forests have significant environmental value, creating a stable ecological environment and promoting biological diversity.

The island of Tasmania, just off the south east coast of Australia has the largest amount of temperate old-growth rainforest reserves in Australia with approximately 1,239,000 hectares in total. While the local Regional Forest Agreement (RFA) was originally designed to protect much of this natural wealth, many of the RFA old growth forests protected in Tasmania consist of trees of little use to the timber industry. RFA old growth and high conservation value forests that contain species highly desirable to the forestry industry have been poorly reserved. Only 22% of Tasmania's original tall-eucalypt forests managed by Forestry Tasmania have been reserved. Ten thousand hectares of tall-eucalypt RFA old growth forest have been lost since 1996, predominantly as a result of industrial logging operations. In 2006, approximately 61,000 hectares of tall-eucalypt RFA old growth forests remained unprotected. Recent logging attempts in the Upper Florentine Valley have sparked a series of protests and media attention over the arrests that have taken place in this area. Additionally, Gunns Limited, the primary forestry contractor in Tasmania has been under recent criticism by political and environmental groups over its practice of woodchipping timber harvested from old growth forests.

Management

The increased understanding of forest dynamics in the late 20th century has led the scientific community to identify a need to inventory, understand, manage and conserve representative examples of old-growth forests with their associated characteristics and values. The literature around old growth and its management is inconclusive about the best way to capture the true essence of an old growth stand.

Old-growth forest in the Opal Creek Wilderness, a wilderness area located in the Willamette National Forest in the U.S. state of Oregon, on the border of the Mount Hood National Forest. It has the largest uncut watershed in Oregon.

A better understanding of natural systems has resulted in new ideas about forest management, such as managed natural disturbances should be designed to achieve the landscape patterns and habitat conditions that are normally maintained in nature (DeLong 1998; Wong and Iverson 2004). This coarse filter approach to biodiversity conservation recognizes ecological processes and provides for a dynamic distribution of old growth across the landscape. And all seral stages – young, medium and old – support forest biodiversity. Plants and animals rely on different forest ecosystem stages to meet their habitat needs.

In Australia, the Regional Forest Agreement (RFA) attempted to prevent the clearfelling of defined "Old Growth Forests". This led to struggles over what constitutes "Old Growth". For example, in Western Australia, the timber industry tried to limit the area of Old Growth in the karri forests of the Southern Forests Region; this led to the creation of the Western Australian Forests Alliance, the splitting of the Liberal Government of Western Australia and the election of the Gallop Labor Government. Old Growth Forests in this region have now been placed inside National Parks. A small proportion of Old Growth Forest also exists in South-West Australia, and is protected by a Federal laws from logging, which hasn't occurred there for more than twenty years.

In British Columbia, Canada, old-growth forests must be maintained in each of the province's ecological units to meet biodiversity needs.

Locations of Remaining Tracts

In 2006 Greenpeace identified that the world's remaining intact forest landscapes are distributed among the continents as follows:

- 35% in Latin America. The Amazon rainforest is mainly located in Brazil, which clears a larger area of forest annually than any other country in the world.
- 28% in North America, which harvests 10,000 square kilometres of ancient forests every year. Many of the fragmented forests of southern Canada and the United States lack adequate

animal travel corridors and functioning ecosystems for large mammals. Most of the remaining old-growth forests in the contiguous United States and Alaska are on public land.

- 19% in Northern Asia, home to the largest boreal forest in the world.
- 8% in Africa, which has lost most of its intact forest landscapes in the last 30 years. The timber industry is responsible for destroying huge areas of intact forest landscapes and continues to be the single largest threat to these areas.
- 7% in South Asia Pacific, where the Paradise Forests are being destroyed faster than any other forest on Earth. Much of the large intact forest landscapes have already been cut down, 72% in Indonesia and 60% in Papua New Guinea.
- Less than 3% in Europe, where more than 150 square kilometres of intact forest landscapes are cleared every year and the last areas of the region's intact forest landscapes in European Russia are shrinking rapidly. In the United Kingdom, they are known as ancient woodlands.

Environmental Audit

Environmental audit is a general term that can reflect various types of evaluations intended to identify environmental compliance and management system implementation gaps, along with related corrective actions. In this way they perform an analogous (similar) function to financial audits. There are generally two different types of environmental audits: compliance audits and management systems audits. Compliance audits tend to be the primary type in the US or within US-based multinationals.

Environmental Compliance Audits

As the name implies, these audits are intended to review the site's/company's legal compliance status in an operational context. Compliance audits generally begin with determining the applicable compliance requirements against which the operations will be assessed. This tends to include federal regulations, state regulations, permits and local ordinances/codes. In some cases, it may also include requirements within legal settlements.

Compliance audits may be *multimedia* or *programmatic*. Multimedia audits involve identifying and auditing all environmental media (air, water, waste, etc.) that apply to the operation/company. Programmatic audits (which may also be called *thematic* or *media-specific*) are limited in scope to pre-identified regulatory areas, such as air.

Audits are also focused on operational aspects of a company/site, rather than the contamination status of the real property. Assessments, studies, etc. that involve property contamination/remediation are typically not considered an environmental audit.

ISO 14001

ISO 14001 is a voluntary international standard for *environmental management systems* ("EMS").

ISO 14001:2004 provides the requirements for an EMS and ISO 14004 gives general EMS guidelines. An EMS meeting the requirements of ISO 14001:2004 is a management tool enabling an organization of any size or type to:

1. Identify and control the environmental impact of its activities, products or services;
2. Improve its environmental performance continually, and
3. Implement a systematic approach to setting environmental objectives and targets, to achieving these and to demonstrating that they have been achieved.

Organizations implementing ISO 14001 usually seek to obtain certification by independent Certification Bodies. Certification indicates that the documentation, implementation and effectiveness of the EMS conform to the specific requirements of ISO 14001. This standard is currently being updated to include elements of including a lifecycle perspective and including top management anongst other changes. The draft (DIS) standard ISODIS 14001:2014 is currently the draft standard applicable until the ISO 14001:2015 standard is finalised and published.

In 2002, the ISO organization also published ISO 19011, the standard for auditing quality and environmental management systems (ISO 19011:2002), which was used for internal audits and certification audits of EMS until it was updated in 2011. The 2011 version on ISO 19011 restricts its use in first and second part audits, while third part audits (certification audits) are now covered in ISO/IEC 17021.

A common misconception is that ISO 14001 certification automatically implies legal compliance. *Certification under ISO 14001 does not directly reflect compliance with any legal requirements*, although ISO 14001 demands the organization to evaluate its compliance with legal requirements. If there is no compliance with some legal requirement, ISO 14001 requires that the organization sets specific targets related to the non-compliance(s) and establishes, implements and maintains programmes to achieve compliance. Therefore it is possible that, at the time of audit, the organization fulfils the requirements of ISO 14001, yet there are one or more non-compliances with specific requirements, which are identified and which the organization actively works to correct. Specific guidance on this subject is provided by the European co-operation for Accreditation.

Audit Tools and Technology

The term "protocol" means the checklist used by environmental auditors as the guide for conducting the audit activities. There is no standard protocol, either in form or content. Typically, companies develop their own protocols to meet their specific compliance requirements and management systems. Audit firms frequently develop general protocols that can be applied to a broad range of companies/operations.

Current technology supports many versions of computer-based protocols that attempt to simplify the audit process by converting regulatory requirements into questions with "yes", "no" and "not applicable" check boxes. Many companies and auditors find these useful and there are several such protocol systems commercially available. Other auditors (typically those with many years of environmental auditing experience) use the regulations/permits directly as protocols. There is a long standing debate among environmental audit professionals on the value of large, highly detailed and prescriptive protocols (i.e., that can, in theory, be completed by an auditor with little or no technical experience) versus more flexible protocols that rely on the expertise and knowledge of experienced auditors and source documents (regulations, permits, etc.) directly. However usage of structured and prescriptive protocols in ISO 14001 audits allows easier review by other parties, either internal to the Certification Body (e.g. technical reviewers and certification managers) or external (accreditation bodies).

In the US, permits for air emissions, wastewater discharges and other operational aspects, many times establish the primary legal compliance standards for companies. In these cases, auditing only to the regulations is inadequate. However, as these permits are site specific, standard protocols are not commercially available that reflect every permit condition for every company/site. Therefore, permit holders and the auditors they hire must identify the permit requirements and determine the most effective way to audit against those requirements.

During the past 20 years, advances in technology have had major impacts on auditing. Laptop computers, portable printers, CD/DVDs, the internet, email and wireless internet access have all been used to improve audits, increase/improve auditor access to regulatory information and create audit reports on-site. At one point in the 1990s, one major company invested significant resources in testing "video audits" where the auditor (located at the corporate headquarters) used real-time video conferencing technology to direct staff at a site to carry live video cameras to specific areas of the plant. While initially promising, this technology/concept did not prove acceptable. An emerging technology in environmental auditing is the use of tablet computers.

Related Types of Assessments

Phase I Environmental Site Assessment ("ESA") are generally done in relation to mergers, acquisitions or financing activities. The intent of ESAs is to identify potential sources/existence of property contamination for purposes of clean up costs/liability under US law. ESA's rarely contain a compliance audit component and should not be confused with audits.

Environmental Auditing in India

The Supreme Audit Institution (SAI) in India is headed by the Comptroller and Auditor General (CAG) of India who is a constitutional authority. The CAG of India derives his mandate from Articles 148 to 151 of the Indian Constitution. The CAG's (Duties, Powers and Conditions of Service) Act, 1971 prescribes functions, duties and powers of the CAG. While fulfilling his constitutional obligations, the CAG examines various aspects of government expenditure and revenues. The audit conducted by CAG is broadly classified into Financial, Compliance and Performance Audit. Environmental audit by SAI India is conducted within the broad framework of compliance and performance audit.

Permissions

All chapters in this book are published with permission under the Creative Commons Attribution Share Alike License or equivalent. Every chapter published in this book has been scrutinized by our experts. Their significance has been extensively debated. The topics covered herein carry significant information for a comprehensive understanding. They may even be implemented as practical applications or may be referred to as a beginning point for further studies.

We would like to thank the editorial team for lending their expertise to make the book truly unique. They have played a crucial role in the development of this book. Without their invaluable contributions this book wouldn't have been possible. They have made vital efforts to compile up to date information on the varied aspects of this subject to make this book a valuable addition to the collection of many professionals and students.

This book was conceptualized with the vision of imparting up-to-date and integrated information in this field. To ensure the same, a matchless editorial board was set up. Every individual on the board went through rigorous rounds of assessment to prove their worth. After which they invested a large part of their time researching and compiling the most relevant data for our readers.

The editorial board has been involved in producing this book since its inception. They have spent rigorous hours researching and exploring the diverse topics which have resulted in the successful publishing of this book. They have passed on their knowledge of decades through this book. To expedite this challenging task, the publisher supported the team at every step. A small team of assistant editors was also appointed to further simplify the editing procedure and attain best results for the readers.

Apart from the editorial board, the designing team has also invested a significant amount of their time in understanding the subject and creating the most relevant covers. They scrutinized every image to scout for the most suitable representation of the subject and create an appropriate cover for the book.

The publishing team has been an ardent support to the editorial, designing and production team. Their endless efforts to recruit the best for this project, has resulted in the accomplishment of this book. They are a veteran in the field of academics and their pool of knowledge is as vast as their experience in printing. Their expertise and guidance has proved useful at every step. Their uncompromising quality standards have made this book an exceptional effort. Their encouragement from time to time has been an inspiration for everyone.

The publisher and the editorial board hope that this book will prove to be a valuable piece of knowledge for students, practitioners and scholars across the globe.

Index

P

R

S

T

W